Oklahoma Herpetology

AN OKLAHOMA MUSEUM OF NATURAL HISTORY PUBLICATION

OKLAHOMA HERPETOLOGY

AN ANNOTATED BIBLIOGRAPHY

Charles C. Carpenter
and James J. Krupa

University of Oklahoma Press · Norman and London

Financial assistance provided by:

The Oklahoma Biological Survey

The Oklahoma Department of Wildlife Conservation's
Nongame Program

The Oklahoma Museum of Natural History

Drawings on the cover and the frontispiece of the collared lizard *(Crotaphytus collaris)* and the Plains spadefoot toad *(Scaphiopus bombifrons),* respectively, are by Michael W. Nickell.

Library of Congress Cataloging-in-Publication Data

Carpenter, Charles C. (Charles Congden), 1921-
Oklahoma herpetology : an annotated bibliography / Charles C. Carpenter and James J. Krupa.
p. cm.
"An Oklahoma Museum of Natural History publication."
Bibliography: p.
Includes index.
ISBN 0-8061-2210-2 (alk. paper)
1. Reptiles—Oklahoma—Bibliography. 2. Amphibians—Oklahoma–Bibliography. I. Krupa, James J., 1956- . II. Title.
Z7996.R4C37 1989
[QL653.O5]
016.5976'09766—dc19 88-38318

First edition.

The paper in this book meets the guidelines for permanence and durability of the Committee on Production Guidelines for Book Longevity of the Council on Library Resources, Inc.

Contents

Preface

Many people have been helpful in the preparation of this publication. We thank Gary D. Schnell, Director of the Oklahoma Biological Survey, for providing financial support to the junior author over the three-year period of this project as well as use of the Biological Survey's computer facilities and personnel. We are especially indebted to Daniel J. Hough, of the Oklahoma Biological Survey, without whose "computer wizardry" and countless hours of effort completion of this publication would have been nearly impossible.

Thanks are extended to Lisa J. Cox, Valerie Y. Guliford, and Jan Ekberg for data entry. Special thanks are extended to Valerie Buckner Visor for doing virtually all the data entry over the final stages of this project.

We also thank George R. Cline, James D. Fawcett, Daniel J. Hough, Geri J. Larkin, Regina H. Macedo, Patrick S. Mulvany, Daniel W. Salzer, Stephen M. Secor, Gary Shugart, and two anonymous reviewers for their comments on various drafts of this manuscript. James D. Fawcett and Gary D. Schnell provided helpful tips on organizing the bibliography. Finally, Don G. Wyckoff, Director of the Oklahoma Archaeological Survey, was helpful in locating the archaeological literature.

The bibliography has been stored and partially processed using GIPSY (General Information Processing System), a general data-handling software package, as implemented at the University of Oklahoma. Considerable additional support software used in processing the bibliography file was created and executed by Daniel J. Hough and Tom M. McKenna. For typesetting, the bibliography was transferred to a personal computer. WordStar 2000+™ software and a Hewlett-Packard LaserJet Series II ™ printer were used to produce the camera-ready copy.

Financial assistance for this project was provided by the Oklahoma Biological Survey, the Oklahoma Museum of Natural History (through Director Michael A. Mares), and the Oklahoma Department of Wildlife Conservation Nongame Wildlife Program (through John Skeen).

Norman, Oklahoma CHARLES C. CARPENTER

JAMES J. KRUPA

Oklahoma Herpetology

Introduction

We have prepared this annotated bibliography so that those interested in Oklahoma herpetology will have easy access to the literature. Great effort has been made to make the bibliography as complete as possible. Although some publications probably were missed, we believe this to be the most thorough compilation to date. In addition to the references in the main body, this bibliography provides the following information: (1) a checklist of Oklahoma's amphibians and reptiles, (2) a Genus and Key Word Index, and (3) an English Name Index.

The bibliography contains references to Oklahoma amphibians and reptiles from the year 1823 (Edwin James's report on Long's Expedition) to December 1987. Regional journals (*Bulletin of the Oklahoma Herpetological Society, Bulletin of the Ornithological Society, Kansas Herpetological Society News Letter, Proceedings of the Oklahoma Academy of Science, Southwestern Naturalist,* and the *Texas Journal of Science*), popular literature (*Outdoor Oklahoma* formerly *Oklahoma Game and Fish News* and *Oklahoma Wildlife*), and North American herpetological journals (*Copeia, Herpetologica, Herpetological Review,* and the *Journal of Herpetology*) were carefully searched. Other publications include: *American Midland Naturalist, American Naturalist, Animal Behaviour, Catalog of American Amphibians and Reptiles, Ecology, Ecological Monographs, Evolution, Great Basin Naturalist, Journal of Mammalogy, Journal of Paleontology, Journal of Parasitology, Systematic Zoology*, and the *Wassman Journal of Biology*. Theses and dissertations from the University of Oklahoma and Oklahoma State University were thoroughly searched. Other helpful sources included: the senior author's extensive reprint collection; *Wildlife Review, Zoological Record*; historical accounts of explorations; wildlife publications; reports from the State Archaeological, Biological, and Geological Surveys; pertinent medical accounts; museum records; books; monographs; and unpublished manuscripts. Of the 1,536 citations in this bibliography, 745 cover amphibians, and 1,267 cover reptiles. A total of 428 publications (28 percent) are from the regional journals, while the major herpetological journals are the source of 257 (17 percent) of the citations.

Each citation was assigned a unique sequence number that is located in the left margin. The citations contain the following information: author, date of publication, title, name of journal or book (the editor is listed if part of a book or symposium), publisher (if not in a periodical), and an annotation. A periodical published by volume and issue (e.g., *Herpetologica*, volume

10, issue 2, pages 3 to 8) appears as "*Herpetologica* 10(2):3-8," while periodicals published once annually lack the issue number in parentheses (e.g., *Proc. Okla. Acad. Sci.* 20:74-77). Occasional papers or miscellaneous publications numbered as part of a series (e.g., *American Museum Novitates,* number 381, pages 1 to 14) are listed in a format similar to periodicals (i.e., *Am. Mus. Nov.* (381):1-14). All periodical abbreviations follow the American National Standard and are summarized in *Serial Sources for the BIOSIS Data Base* (Bioscience Information Service, 2100 Arch Street, Philadelphia, Pennsylvania 19103).

The annotation often begins with a short description when the title fails to convey the pertinence of the article. It then lists relevant Key Words. The following Key Words were used: archeology, behavior, bibliography, development, distribution, ecology, economics, evolution, general reference, genetics, health, historical, husbandry, lore, management, morphology, natural history, paleontology, parasitology, physiology, popular, taxonomy, and techniques. The annotation ends with a list of genera appearing in the article (Table I).

When possible, collective phrases (see Table I) replace genus names to reduce the length of annotations. For example, if all seven genera of Oklahoma frogs and toads occur in an article, then the phrase "All Anura" is used. Similarly, "All Caudata" is used in place of the ten genera of salamanders. Furthermore, "All Amphibia" indicates that the eighteen genera of Oklahoma amphibians appeared. The same convention is used for reptiles. "All Testudines," "All Sauria," "All Serpentes," and "All Reptilia" indicate all genera of turtles, lizards, snakes, and reptiles, respectively, that appeared in the article. Some articles deal with extinct forms or make general reference to certain extant groups without indicating species. For these articles, the phrase "General Reference" appears in the annotation.

The organization of this bibliography is intended for easy retrieval of information. The reader can go directly to the main body of the bibliography to find an author's publications and the contents of each article. On the other hand, the Genus and Key Word Index allows the reader to find articles for a genus; under the generic designation, we have categorized articles by subject. For example, under the genus *Bufo*, a series of sequence numbers are listed by subject such as Behavior or Taxonomy. By referring back to these numbers in the bibliography, the reader will find the complete citation. Similarly, those articles categorized as "General Reference" are listed by subject at the end of the index.

We hope that the reader has easy access to the needed information found in this bibliography.

Table I. *The seven collective "All" phrases and the sixty-four genera used for key words appear below.*

All Amphibia		All Reptilia			
All Anura	All Caudata	All Testudines	All Sauria	All Serpentes	Crocodilia
Bufo	*Ambystoma*	*Chelydra*	*Ophisaurus*	*Arizona*	*Alligator*
Acris	*Amphiuma*	*Macroclemys*	*Anolis*	*Carphophis*	
Hyla	*Desmognathus*	*Chrysemys*	*Crotaphytus*	*Cemophora*	
Pseudacris	*Eurycea*	*Deirochelys*	*Holbrookia*	*Coluber*	
Gastrophryne	*Hemidactylum*	*Graptemys*	*Phrynosoma*	*Diadophis*	
Scaphiopus	*Plethodon*	*Pseudemys*	*Sceloporus*	*Elaphe*	
Rana	*Typhlotriton*	*Terrapene*	*Uta*	*Farancia*	
	Necturus	*Trachemys*	*Eumeces*	*Heterodon*	
	Siren	*Kinosternon*	*Scincella*	*Hypsiglena*	
	Notophthalmus	*Sternotherus*	*Cnemidophorus*	*Lampropeltis*	
		Trionyx		*Masticophis*	
				Nerodia	
				Opheodrys	
				Pituophis	
				Regina	
				Rhinocheilus	
				Sonora	
				Storeria	
				Tantilla	
				Thamnophis	
				Tropidoclonion	
				Virginia	
				Leptotyphlops	
				Agkistrodon	
				Crotalus	
				Sistrurus	

Checklist of Extant Oklahoma Amphibians and Reptiles

Checklist of Extant Oklahoma Amphibians and Reptiles

Scientific names and common names follow Banks et al. (1987; Checklist of Vertebrates of the U.S., U.S. Territories, and Canada; U.S. Dept. Interior Fish Wildl. Serv. Resourse Publ. 166).

ORDER ANURA (Frogs and Toads)

BUFONIDAE (Bufonids and Toads)

Bufo americanus (American toad)
Bufo cognatus (Great Plains toad)
Bufo debilis (Green toad)
Bufo punctatus (Red-spotted toad)
Bufo speciosus (Texas toad)
Bufo woodhousii (Woodhouse's toad)

HYLIDAE (Hylids and Treefrogs)

Acris crepitans (Northern cricket frog)
Hyla avivoca (Bird-voiced treefrog)
Hyla cinerea (Green treefrog)
Hyla chrysoscelis (Cope's gray treefrog)
Hyla crucifer (Spring peeper)
Hyla versicolor (Gray treefrog)
Pseudacris clarkii (Spotted chorus frog)
Pseudacris streckeri (Strecker's chorus frog)
Pseudacris triseriata (Striped chorus frog)

MICROHYLIDAE (Microhylids)

Gastrophryne carolinensis (Eastern narrow-mouthed toad)
Gastrophryne olivacea (Western narrow-mouthed toad)

PELOBATIDAE (Pelobatids and Spadefoots)

Scaphiopus bombifrons (Plains spadefoot)
Scaphiopus couchii (Couch's spadefoot)
Scaphiopus holbrookii (Eastern spadefoot)
Scaphiopus hammondii (Western spadefoot)

RANIDAE (Ranids)

Rana areolata (Crawfish frog)
Rana blairi (Plains leopard frog)

RANIDAE *(continued)*
Rana catesbeiana (Bullfrog)
Rana clamitans (Green frog)
Rana palustris (Pickerel frog)
Rana pipiens (Northern leopard frog)
Rana sphenocephala (Southern leopard frog)

ORDER CAUDATA (Salamanders)

AMBYSTOMATIDAE (Mole Salamanders)

Ambystoma annulatum (Ringed salamander)
Ambystoma maculatum (Spotted salamander)
Ambystoma opacum (Marbled salamander)
Ambystoma talpoideum (Mole salamander)
Ambystoma texanum (Smallmouth salamander)
Ambystoma tigrinum (Tiger salamander)

AMPHIUMIDAE (Amphiumas)

Amphiuma tridactylum (Three-toed amphiuma)

PLETHODONTIDAE (Lungless Salamanders)

Desmognathus brimleyorum (Ouachita dusky salamander)
Eurycea longicauda (Long-tailed salamander)
Eurycea lucifuga (Cave salamander)
Eurycea multiplicata (Many-ribbed salamander)
Eurycea quadridigitata (Dwarf salamander)
Eurycea tynerensis (Oklahoma salamander)
Hemidactylium scutatum (Four-toed salamander)
Plethodon dorsalis (Zigzag salamander)
Plethodon glutinosus (Slimy salamander)
Plethodon ouachitae (Rich Mountain salamander)
Plethodon serratus (Southern red-backed salamander)
Typhlotriton spelaeus (Grotto salamander)

PROTEIDAE (Mudpuppies and Waterdogs)

Necturus maculosus (Mudpuppy)

SIRENIDAE (Sirens)

Siren intermedia (Lesser siren)

SALAMANDRIDAE (Newts and Salamanders)

Notophthalmus viridescens (Eastern newt)

ORDER TESTUDINES (Turtles)

CHELYDRIDAE (Snapping Turtles)

Chelydra serpentina (Snapping turtle)
Macroclemys temminckii (Alligator snapping turtle)

EMYDIDAE (Emydids)

Chrysemys picta (Painted turtle)
Deirochelys reticularia (Chicken turtle)
Graptemys geographica (Map turtle)
Graptemys kohnii (Mississippi map turtle)
Graptemys ouachitensis (Ouachita map turtle)
Graptemys pseudogeographica (False map turtle)
Pseudemys concinna (Eastern cooter)
Terrapene carolina (Eastern box turtle)
Terrapene ornata (Ornate box turtle)
Trachemys scripta (Common slider)

KINOSTERNIDAE (Mud and Musk Turtles)

Kinosternon flavescens (Yellow mud turtle)
Kinosternon subrubrum (Common mud turtle)
Sternotherus carinatus (Razor-backed musk turtle)
Sternotherus odoratus (Common musk turtle)

TRIONYCHIDAE (Softshell Turtles)

Trionyx muticus (Smooth softshell turtle)
Trionyx spiniferus (Spiny softshell turtle)

ORDER SQUAMATA (Lizards and Snakes)

SUBORDER SAURIA (Lizards)

ANGUIDAE (Anguids, Alligator, and Glass Lizards)

Ophisaurus attenuatus (Slender glass lizard)

IGUANIDAE (Iguanids)

Anolis carolinensis (Green anole)
Crotaphytus collaris (Collared lizard)
Holbrookia maculata (Lesser earless lizard)
Phrynosoma cornutum (Texas horned lizard)
Phrynosoma modestum (Round-tailed horned lizard)
Sceloporus olivaceus (Texas spiny lizard)
Sceloporus undulatus (Eastern fence lizard)

IGUANIDAE *(continued)*

Uta stansburiana (Side-blotched lizard)

SCINCIDAE (Skinks)

Eumeces anthracinus (Coal skink)
Eumeces fasciatus (Five-lined skink)
Eumeces laticeps (Broad-headed skink)
Eumeces obsoletus (Great Plains skink)
Eumeces septentrionalis (Prairie skink)
Scincella lateralis (Ground skink)

TEIIDAE (Whiptails and Racerunners)

Cnemidophorus gularis (Texas spotted whiptail)
Cnemidophorus sexlineatus (Six-lined racerunner)
Cnemidophorus tesselatus (Checkered whiptail)

ORDER SQUAMATA (Lizards and Snakes)
SUBORDER SERPENTES (Snakes)

COLUBRIDAE (Colubrids)

Arizona elegans (Glossy snake)
Carphophis amoenus (Worm snake)
Cemophora coccinea (Scarlet snake)
Coluber constrictor (Racer)
Diadophis punctatus (Ring-necked snake)
Elaphe guttata (Corn snake)
Elaphe obsoleta (Rat snake)
Farancia abacura (Mud snake)
Heterodon nasicus (Western hog-nosed snake)
Heterodon platyrhinos (Eastern hog-nosed snake)
Hypsiglena torquata (Night snake)
Lampropeltis calligaster (Prairie kingsnake)
Lampropeltis getulus (Common kingsnake)
Lampropeltis triangulum (Milk snake)
Masticophis flagellum (Coachwhip)
Nerodia erythrogaster (Plain-bellied water snake)
Nerodia fasciata (Southern water snake)
Nerodia rhombifera (Diamondback water snake)
Nerodia sipedon (Northern water snake)
Opheodrys aestivus (Rough green snake)
Pituophis melanoleucus (Bull snake)
Regina grahamii (Graham's crayfish snake)
Regina rigida (Glossy crayfish snake)

COLUBRIDAE (*continued*)

Rhinocheilus lecontei (Long-nosed snake)
Sonora semiannulata (Ground snake)
Storeria dekayi (Brown snake)
Storeria occipitomaculata (Red-bellied snake)
Tantilla gracilis (Flat-headed snake)
Tantilla nigriceps (Plains black-headed snake)
Thamnophis cyrtopsis (Black-necked garter snake)
Thamnophis elegans (Western terrestrial garter snake)
Thamnophis marcianus (Checkered garter snake)
Thamnophis proximus (Western ribbon snake)
Thamnophis radix (Plains garter snake)
Thamnophis sirtalis (Common garter snake)
Tropidoclonion lineatum (Lined snake)
Virginia striatula (Rough earth snake)
Virginia valeriae (Smooth earth snake)

LEPTOTYPHLOPIDAE (Slender Blind Snakes)

Leptotyphlops dulcis (Texas slender blind snake)

VIPERIDAE (Pit Vipers)

Agkistrodon contortrix (Copperhead)
Agkistrodon piscivorus (Cottonmouth)
Crotalus atrox (Western diamondback rattlesnake)
Crotalus horridus (Timber rattlesnake)
Crotalus viridis (Western rattlesnake)
Sistrurus catenatus (Massasauga)
Sistrurus miliarius (Pigmy rattlesnake)

ORDER CROCODILIA

ALLIGATORIDAE (Alligators)

Alligator mississippiensis (American alligator)

Annotated Bibliography

Annotated Bibliography

Acker, G., *see* 1316

1 **Adler, K. K.** 1968. Synonymy of the Pliocene turtles *Pseudemys hilli* Cope and *Chrysemys limnodytes* Galbreath. J. Herpetol. 1(1-4):32-38.—Paleontology. *Chrysemys.*

2 **Alciatore, M. L.** 1969. The relationship of the position of the visceral organs to the ventral scale number in three species of snakes. M.S. Thesis, Okla. State Univ., Stillwater. 30 pp.—Morphology. *Nerodia, Virginia, Agkistrodon.*

3 **Aldrich, J. W., and C. G. Endicott.** 1984. Black rat snake predation on giant Canada goose eggs. Wildl. Soc. Bull. 12(3):263-264.—Natural History. *Elaphe.*

4 **Allen, R. F.** 1961. Histological and biochemical studies on *Crotalus atrox.* M.S. Thesis, Univ. Okla., Norman. 54 pp.—Physiology. *Crotalus.*

5 **Allen, R. F., L. G. Gumbreck, and M. R. Shetlar.** 1959. Some preliminary studies of "stress" in the western diamondback rattlesnake (*Crotalus atrox*). Proc. Okla. Acad. Sci. 40:19-21.—Parasitology. *Crotalus.*

6 **Allen, R. F., and J. F. Lhotka, Jr.** 1982. *Crotalus atrox:* notes on the gastric wall. Copeia 1982(1):198-200.—Morphology. *Crotalus.*

7 **Altig, R.** 1970. A key to the tadpoles of the continental United States and Canada. Herpetologica 26(2):180-207.—Bibliographies, Morphology. All Anura.

8 **Altig, R., and P. H. Ireland.** 1984. A key to salamander larvae and larviform adults of the United States and Canada. Herpetologica 40(2):212-218.—Taxonomy. All Caudata.

9 **Altig, R., and R. Lohoefener.** 1983. *Rana areolata.* Cat. Am. Amphib. Reptiles 324.1-324.4.—Range map and literature review. Bibliographies, Distribution. *Rana.*

Altig, R., *see* 821

10 **Altman, R. W., and G. R. Cline.** 1985. Introduction to the snakes of Oklahoma. Coop. Ext. Serv. Div. Arg. Okla. State Univ. (9010):1-6.—Natural History, Popular. All Serpentes.

11 **Amaral, A. D.** 1929. Studies of Nearctic Ophidia V. on *Crotalus confluentus* Say, 1823, and its allied forms. Bull. Antivenin Inst. Am. 2(4):86-97.—Distribution. *Crotalus.*

12 **Anderson, J. D.** 1965. *Ambystoma annulatum.* Cat. Am. Amphib. Reptiles 19.1-19.2.—Range map and literature review. Bibliographies, Distribution. *Ambystoma.*

13 **Anderson, J. D.** 1967. *Ambystoma maculatum.* Cat. Am. Amphib. Reptiles 51.1-51.4.—Range map and literature review. Bibliographies, Distribution. *Ambystoma.*

14 **Anderson, J. D.** 1967. *Ambystoma opacum.* Cat. Am. Amphib. Reptiles 46.1-46.2.—Range map and literature review. Bibliographies, Distribution. *Ambystoma.*

15 **Anderson, J. D.** 1967. *Ambystoma texanum.* Cat. Am. Amphib. Reptiles 37.1-37.2.—Range map and literature review. Bibliographies, Distribution. *Ambystoma.*

16 **Andrews, L. E.** 1981. Field trip to the Gypsum Hills area of Kingfisher County, Oklahoma. Bull. Okla. Herp. Soc. 6(2):48.—Author relives the joy of a collecting trip. Distribution, Popular. *Crotaphytus, Diadophis, Sonora.*

17 **Anonymous.** 1930. Snake stories. Lit. Dig. 106(6):21.—General Reference, Lore, Popular.

18 **Anonymous.** 1946. Deep fork fishermen capture 100 pound turtle. Okla. Game Fish News 2(8):16.—Photograph included. Popular. *Macroclemys.*

19 **Anonymous.** 1946. Deep fork rattler. Okla. Game Fish News 2(10):20.—Photograph of timber rattlesnake included. Popular. *Crotalus.*

20 **Anonymous.** 1946. Mammoth turtle captured in Blue River. Okla. Game Fish News 2(2):20.—Photograph included. Popular. *Macroclemys.*

21 **Anonymous.** 1946. Moccasin shooter. Okla. Game Fish News 2(5):18.—Photograph of water moccasins. Popular. *Agkistrodon.*

22 **Anonymous.** 1946. Rattlesnakes in pairs. Okla. Game Fish News 2(8):18.—Photographs of rattlesnakes. Popular. *Crotalus.*

23 **Anonymous.** 1946. Snake story. Okla. Game Fish News 2(9):22.—Story of a rattlesnake that ate a squirrel. Popular. *Crotalus.*

24 **Anonymous.** 1946. 50 gallons of soup, on the hoof. Okla. Game Fish News 2(4):22.—Photograph included. Popular. *Macroclemys.*

25 **Anonymous.** 1947. Conservation goes to the fair. Okla. Game Fish News 3(11):12-13.—Photograph. Popular. *Macroclemys.*

26 **Anonymous.** 1947. Giant turtles captured in eastern Oklahoma streams. Okla. Game Fish News 3(7):11.—Photographs included. Popular. *Macroclemys.*

27 **Anonymous.** 1950. Season's first big snapper. Okla. Game Fish News 6(7-8):22.—Photograph. Popular. *Macroclemys.*

28 **Anonymous.** 1951. Okla. Game Fish News 7(5):12.—Photograph. Popular. *Macroclemys.*

29 **Anonymous.** 1952. Dramatic dining. Okla. Game Fish News 8(11): 20.—Popular. *Pituophis.*

30 **Anonymous.** 1952. 300 years old? Okla. Game Fish News 8(9):6.—Photograph of a 120 pound snapping turtle from Miami, Oklahoma. Popular. *Macroclemys.*

31 **Anonymous.** 1953. Hobby: Turtles. Okla. Game Fish News 9(1): 22.—Popular. *Chelydra, Trachemys.*

32 **Anonymous.** 1954. Okla. Game Fish News 10(12):8.—Photograph of a dead diamondback rattlesnake. Popular. *Crotalus.*

33 **Anonymous.** 1954. King size. Okla. Game Fish News 10(10):11.—Caption and photograph of bullfrogs collected in a hunt. Popular. *Rana.*

34 **Anonymous.** 1954. Rattlesnake roundup. Okla. Game Fish News 10(4):5.—Popular. *Crotalus.*

35 **Anonymous.** 1954. Snakes alive. Okla. Game Fish News 10(9):10.—Popular. *Heterodon.*

36 **Anonymous.** 1955. Okla. Game Fish News 11(7-8):7.—Photographs of three lizards. Popular. *Crotaphytus, Phrynosoma, Eumeces.*

37 **Anonymous.** 1955. Big one that didn't get away! Okla. Game Fish News 11(3):15.—Popular. *Macroclemys.*

38 **Anonymous.** 1955. The bullfrog. Okla. Game Fish News 11(7-8):3.—Incorrectly printed as volume 10. Popular. *Rana.*

39 **Anonymous.** 1955. Huge snapping turtle creates interest. Okla. Game Fish News 11(12):10.—Incorrectly printed as issue 11. Popular. *Macroclemys.*

40 **Anonymous.** 1956. Season's largest turtle? Okla. Game Fish News 12(7):10.—Popular. *Macroclemys.*

41 **Anonymous.** 1958. Okla. Wildl. 14(9):9.—Photograph of a diamondback rattlesnake. Popular. *Crotalus.*

42 **Anonymous.** 1958. "Bull-frogging" is popular pastime. Okla. Wildl. 14(7-8):70.—Photograph of bullfrogs killed in a hunt. Popular. *Rana.*

43 **Anonymous.** 1959. Okla. Wildl. 15(12):10.—Photograph of dead water snake and its 51 young. Popular. *Nerodia.*

44 **Anonymous.** 1960. Okla. Wildl. 16(9):13.—Photograph of a snapping turtle. Popular. *Macroclemys.*

45 **Anonymous.** 1960. Okla. Wildl. 16(9):13.—Photograph of common snapping turtle. Popular. *Chelydra.*

46 **Anonymous.** 1960. Okla. Wildl. 16(7-8):20.—Photograph of 7 water moccasins killed for sport. Incorrectly printed as issue 6. Popular. *Agkistrodon.*

47 **Anonymous.** 1960. Cover photo. Okla. Wildl. 16(9):2.—Photograph of bullfrog and brief mention of some anuran calls heard in Oklahoma. Popular. *Bufo, Hyla, Pseudacris, Rana.*

48 **Anonymous.** 1961. Okla. Wildl. 17(2):20.—Photograph of coachwhip snake swallowing a cottonrat. Popular. *Masticophis.*

49 **Anonymous.** 1962. Okla. Wildl. 18(7-8):8.—Photograph of a snapping turtle from Salt Creek. Popular. *Macroclemys.*

50 **Anonymous.** 1962. Loose jointed wiggler. Okla. Wildl. 18(12):9.—Incorrectly printed as volume 19, issue 11. Popular. *Ophisaurus.*

51 **Anonymous.** 1962. Snake Hunt. Okla. Wildl. 18(7-8):9.—164 cottonmouths and copperheads slaughtered in first annual cottonmouth hunt. Popular. *Agkistrodon.*

52 **Anonymous.** 1963. Nature's theatre of knowledge. Okla. Wildl. 19(6):4.—Incorrectly printed as volume 20. Popular. *Sceloporus.*

53 **Anonymous.** 1963. Rattling good frog study? Okla. Wildl. 19(6): 20.—Photograph shows prairie rattlesnake removed from bullfrog stomach. Incorrectly printed as volume 20. Popular. *Rana, Crotalus.*

54 **Anonymous.** 1964. Okla. Wildl. 20(9):17.—Photograph showing Dr. Wyche who caught a 3.5 foot alligator while fishing in Lake Schooler. Popular. *Alligator.*

55 **Anonymous.** 1964. Wildlife mail bag. Okla. Wildl. 20(3):21-22. —Photograph of "spotted" salamander on road. Popular. *Ambystoma.*

56 **Anonymous.** 1965. Camera roundup. Okla. Wildl. 21(6):20-21.—Photograph of diamondback rattlesnake from Waurika roundup. Popular. *Crotalus.*

57 **Anonymous.** 1965. Wildlife parade. Okla. Wildl. 21(6):16-17.—Report of a big turtle that swallowed a beaver. Lore, Popular. *Chelydra.*

58 **Anonymous.** 1976. Wildlife portrait: The toad. Outdoor Okla. 32(7): 24.—Photograph included inside front cover. Popular. *Bufo.*

59 **Anonymous.** 1977. The pond. Outdoor Okla. 33(8):5-9.—Photographs are included. Popular. *Acris, Trachemys.*

60 **Anonymous.** 1978. Wildlife portrait: The box turtle. Outdoor Okla. 34(4):20.—Photograph included inside front cover. Popular. *Terrapene.*

61 **Anonymous.** 1980. Wildlife portrait: The copperhead. Outdoor Okla. 36(4):20.—Photograph included inside front cover. Popular. *Agkistrodon.*

62 **Anonymous.** 1981. Opinion: The endangered rattler. Bull. Okla. Herp. Soc. 6(2):57.—Economics, Lore. *Crotalus, Sistrurus.*

63 **Anonymous.** 1981. Wildlife portrait: The bullfrog. Outdoor Okla. 37(3):56.—Popular. *Rana.*

64 **Anonymous.** 1982. Wildlife laws of Oklahoma. Oklahoma Statutes, Title 29.ss 1111 through 1113, As amended through laws of 1982 regular session of the Legislature.—Economics, Management. *Rana.*

65 **Anonymous.** 1983. Wildlife portrait: The toad. Outdoor Okla. 39(2): 56-57.—Popular. *Bufo.*

66 **Anonymous.** 1985. Threatened and endangered species. Southest-Cent. Okla. Study Environ. Res. Inven. N.U.S. Corporation. 151 pp.—A study prepared for the U.S. Department of the Interior's Bureau of Reclamation. Distribution, Management. *Eurycea, Plethodon, Alligator.*

67 **Anonymous.** 1986. Spiro mounds exhibit. Stovall Mus. Newsl. 10(4): 2.—Line drawing of entwined snakes. General Reference, Lore.

68 **Appley, M. B.** 1967. The distribution of the outgrowth orienting factor (OOF) among various vertebrate groups. M.S. Thesis, Univ. Okla., Norman. 60 pp.—Physiology. *Crotalus.*

69 **Ashton, R. E., Jr.** 1976. Endangered and threatened amphibians and reptiles in the United States. Soc. Study Amphib. Reptiles Herpetol. Circ. (5):1-65.—Management. *Hyla, Scaphiopus, Ambystoma, Amphiuma, Eurycea, Hemidactylium, Plethodon, Typhlotriton, Chrysemys, Graptemys, Phrynosoma, Sceloporus, Uta, Cnemidophorus, Cemophora, Farancia, Regina, Thamnophis, Alligator.*

70 **Auffenberg, W.** 1955. A reconsideration of the racer, *Coluber constrictor,* in eastern United States. Tulane Stud. Zool. 2(6):89-155.—Distribution, Evolution, Taxonomy. *Coluber.*

71 **Auffenberg, W., and W. W. Milstead.** 1965. Reptiles of the Quaternary of North America. Pages 557-568 *in* H. E. Wright Jr. and D. G. Frey (Eds.). The Quaternary of the United States. Princeton Univ. Press, Princeton. 922 pp.—Distribution, Evolution, Paleontology. *Chrysemys, Trachemys, Kinosternon, Trionyx, Ophisaurus, Eumeces, Coluber, Elaphe, Lampropeltis, Opheodrys, Pituophis.*

Ault, III, J. W., *see* 1441

72 **Axtell, R. W.** 1956. A solution to the long neglected *Holbrookia lacerata* problem, and the description of two new subspecies of *Holbrookia.* Bull. Chic. Acad. Sci. 10(11):163-179.—Behavior, Distribution, Morphology, Natural History, Taxonomy. *Holbrookia.*

73 **Axtell, R. W.** 1963. A reinterpretation of the distribution of *Bufo w. woodhousei* Girard, especially on the southeastern margin of its range. Herpetologica 19(2):115-122.—Distribution. *Bufo.*

74 **Axtell, R. W.** 1983. Range portrayal and reality: *Heterodon platyrhinos* distribution on the High Plains of Texas and Oklahoma. J. Herpetol. 17(2):191-193.—Distribution. *Heterodon.*

75 **Axtell, R. W.** 1987. More accurate range portrayal, even on smaller maps? Axtell's answer to Blem's protestations. Herpetol. Rev. 18(1): 9-10.—Distribution, Techniques. *Heterodon.*

76 **Bailey, V.** 1905. Biological Survey of Texas. North Am. Fauna (25): 9-222.—Distribution. *Crotalus.*

77 **Baird, S. F.** 1859. In General report upon the zoology of the several Pacific Railroad routes. Sen. Exec. Doc. 78. 2nd. Sess. 33rd. Cong. 10(3-4):14-45.—Distribution, Historical, Taxonomy. *Bufo, Acris, Rana, Necturus, Crotaphytus, Sceloporus, Elaphe, Nerodia, Opheodrys, Pituophis, Thamnophis, Agkistrodon, Crotalus, Sistrurus.*

78 **Baird, S. F.** 1859. Reptiles of the boundary. *In* W. H. Emory. Report on the United States and Mexican Boundary Survey. Sen. Exec. Doc. 108. 1st. Sess. 34th. Cong. 35 pp.—Distribution. General Reference.

79 **Baird, S. F., and C. Girard.** 1853. Catalogue of North American reptiles in the museum of the Smithsonian Institution. Part I. Serpents. Publ. Smithson. Inst. 172 pp.—Distribution, Taxonomy. *Elaphe, Heterodon, Lampropeltis, Masticophis, Nerodia, Opheodrys, Pituophis, Thamnophis, Crotalus.*

80 **Baird, S. F., and C. Girard.** 1853. Reptiles. Pages 188-215 *in* R. B. Marcy and G. B. McClellan. (Eds.). Exploration of the Red River of Louisiana in the year 1852. Sen. Exec. Doc. 1st Sess. 33rd Cong. 286 pp.—Distribution, Historical. *Bufo, Rana, Crotaphytus, Holbrookia, Phrynosoma, Sceloporus, Cnemidophorus, Scincella, Lampropeltis, Opheodrys, Thamnophis, Crotalus.*

81 **Baker, R. E.** 1983. The western cottonmouth in Osage County, Oklahoma. Bull. Okla. Herp. Soc. 8(2):56-57.—Distribution. *Agkistrodon.*

82 **Baker, R. E.** 1985. Collared lizard wards off a young opossum. Bull. Okla. Herp. Soc. 10(3-4):28.—Behavior. *Crotaphytus.*

83 **Baker, R. E.** 1985. Late season capture of the western cottonmouth, *Agkistrodon piscivorus leucostoma,* in northern Osage County, Oklahoma. Bull. Okla. Herp. Soc. 10(3-4):23.—Natural History. *Agkistrodon.*

Baker, R. E., *see* 684

84 **Baker, R. J., G. A. Mengden, and J. J. Bull.** 1972. Karyotypic studies of thirty-eight species of North American snakes. Copeia 1972(2): 257-265.—Genetics, Morphology. *Agkistrodon.*

Bakker, R. T., *see* 580

85 **Baldauf, R. J.** 1955. Contributions to the cranial morphology of *Bufo w. woodhousei* Girard. Tex. J. Sci. 7(3):275-311.—Some of the specimens were from the University of Oklahoma Museum of Zoology. Morphology. *Bufo.*

86 **Ball, R. L.** 1975. Geographic distribution: *Cemophora coccinea copei.* Herpetol. Rev. 6(2):45.—Distribution. *Cemophora.*

87 **Ball, R. L.** 1975. Geographic distribution: *Sistrurus miliarius streckeri.* Herpetol. Rev. 6(2):45.—Distribution. *Sistrurus.*

88 **Ball, R. L.** 1979. Aquatic foraging behavior of the black rat snake. Bull. Okla. Herp. Soc. 4(2-3):38-39.—Behavior, Natural History. *Elaphe.*

89 **Ball, R. L.** 1979. Communal egg laying of the black rat snake in central Oklahoma. Bull. Okla. Herp. Soc. 4(2-3):33-34.—Natural History. *Elaphe.*

90 **Ball, R. L.** 1979. Geographic distribution: *Cemophora coccinea.* Herpetol. Rev. 10(4):118.—Distribution. *Cemophora.*

91 **Ball, R. L.** 1979. A new county record of the northern scarlet snake in central Oklahoma. Bull. Okla. Herp. Soc. 4(2-3):45.—Distribution. *Cemophora.*

Barbour, R. W., *see* 617

Barbour, T., *see* 1355, 1356, 1357, 1358, 1359

Barghusen, H., *see* 1135

92 **Barkas, J., J. Benefield, and R. D. Grimpe.** 1979. Anomalies in a Rocky Mountain toad. Bull. Okla. Herp. Soc. 4(2-3):37.—Morphology. *Bufo.*

93 **Barkley, F. A.** 1977. History of Oklahoma Herpetology III. The Oklahoma University, Department of Zoology and the Zoological Survey, Summer of 1926. Bull. Okla. Herp. Soc. 2(2-3):56-63.—Historical. *Scaphiopus, Rana, Terrapene, Crotaphytus, Eumeces, Rhinocheilus, Leptotyphlops, Crotalus.*

Barkley, F. A., *see* 150

94 **Barnes, S. B., J. H. Black, and J. Pigg.** 1986. New distribution records of the genus *Sternotherus* in Oklahoma. Bull. Chic. Herp. Soc. 21(3-4):98-99.—Distribution. *Sternotherus.*

95 **Bart, H. L., and J. J. Hoover.** 1984. Delaware County record for the western cottonmouth. Bull. Okla. Herp. Soc. 9(4):35.—Distribution. *Agkistrodon.*

Baugh, T. G., *see* 585

96 **Baur, G.** 1893. Notes on the classification and taxonomy of the Testudinidae. IV. The species of the genus *Pseudemys.* Proc. Amer. Philos. Soc. 31:221-225.—Taxonomy. *Pseudemys, Trachemys.*

97 **Beal, K. G.** 1981. Winter foraging habits of the roadrunner. Bull. Okla. Ornithol. Soc. 14(1):13-15.—Lizard and snake mentioned as possible food items. General Reference, Natural History.

98 **Beck, M. L., and J. T. Mahan.** 1979. Ammoniacal silver staining of nucleolar organizer regions in four species of *Bufo.* Copeia 1979(2): 341-345.—Woodhouse toads were collected in Norman. Figure shows karyotype of this species. Genetics. *Bufo.*

99 **Becker, J.** 1979. Frogging for fun. Outdoor Okla. 35(6):15-17.—Popular. *Rana.*

100 **Becker, J.** 1980. Natural baits. Outdoor Okla. 36(6):6-9.—Included the use of tadpoles as bait. General Reference, Natural History.

101 **Behler, J. L., and F. W. King.** 1979. The Audubon Society field guide to North American reptiles and amphibians. Chanticleer Press, Inc., New York. 719 pp.—Distribution, Morphology, Natural History, Popular, Taxonomy, All Amphibia, All Reptilia.

Behrensmeyer, A. K., *see* 580

102 **Bell, J. R.** 1973. The journal of Captain John R. Bell, Official Journalist for the Stephen H. Long Expedition to the Rocky Mountains, 1820. *In* L. R. Hafen (Ed.). Arthur H. Clark Co., Glendale, California.—Historical. *Kinosternon, Crotalus.*

103 **Bellis, E. D.** 1953. The effects of temperature on the breeding calls of some Oklahoma salientians. M.S. Thesis, Univ. Okla., Norman. 36 pp.—Behavior. *Bufo, Hyla, Pseudacris.*

104 **Bellis, E. D.** 1957. The effects of temperature on salientian breeding calls. Copeia 1957(2):85-89.—Behavior, Physiology. *Bufo, Pseudacris.*

Benefield, G. E., *see* 727

Benefield, J., *see* 92, 728

105 **Bennett, C., and R. J. Taylor.** 1968. Notes on the lesser siren, *Siren intermedia* (Urodela). Southwest. Nat. 13(4):455-457.—Twenty-five specimens were electro-shocked and rotenoned out of Raymond Gary Lake, Choctaw County. Distribution, Techniques. *Siren.*

Bennett, W. J., Jr., *see* 1187

Bickham, J. W., *see* 792

106 **Bigham, S. R., L. Hepworth, and R. Martin.** 1964. A casualty count of wildlife following a fire. Proc. Okla. Acad. Sci. 45:47-50.—Ecology. *Terrapene, Elaphe, Masticophis, Agkistrodon, Crotalus.*

107 **Bigham, S. R., and R. L. Lardie.** 1976. Albino ornate box turtle from Woods County, Oklahoma. Bull. Okla. Herp. Soc. 1(2):24.—Morphology. *Terrapene.*

Bird, R. D., *see* 1158

108 **Bishop, S. C.** 1943. Handbook of salamanders: The salamanders of the United States, of Canada, and of lower California. Comstock Publ. Co., Ithaca. 555 pp.—Distribution. Natural History, *Ambystoma, Desmognathus, Eurycea, Plethodon, Typhlotriton, Necturus, Siren, Notophthalmus.*

109 **Bishop, S. C.** 1944. A new neotenic plethodont salamander, with notes on related species. Copeia 1944(1):1-5.—Article includes collection locations in Oklahoma. Distribution, Taxonomy. *Typhlotriton.*

110 **Black, A. H.** 1987. Natural history and food habits of the ground skink in two Oklahoma counties. Bull. Okla. Herp. Soc. 12(1-4):7-12.—Ecology, Natural History. *Scincella.*

Black, A. H., *see* 148, 149, 150, 161

Black, C. C., *see* 867

111 **Black, J. H.** 1968. A possible stimulus for the formation of some aggregations in tadpoles of *Scaphiopus bombifrons.* Proc. Okla. Acad. Sci. 49:13-14.—Behavior, Natural History. *Scaphiopus.*

112 **Black, J. H.** 1969. A cave dwelling population of *Ambystoma tigrinum mavortium* in Oklahoma. J. Herpetol. 3(3-4):183-184.—Ecology. *Ambystoma.*

113 **Black, J. H.** 1969. Ethoecology of tadpoles of *Bufo americanus charlesmithi* in a temporary pool in Oklahoma. J. Herpetol. 3(3-4): 198.—Behavior, Ecology, Natural History. *Bufo.*

114 **Black, J. H.** 1969. Notes on tiger salamanders. Okla. Underground 2(2):48-50.—Natural History. *Ambystoma.*

115 **Black, J. H.** 1971. Boat-tail grackles feeding tadpoles to nestlings. Bull. Okla. Ornithol. Soc. 3(4):33-34.—Natural History. *Scaphiopus.*

116 **Black, J. H.** 1971. The cave life of Oklahoma: A preliminary study excluding Chiroptera. Okla. Underground 4(1-2):2-53.—Distribution, Ecology. *Bufo, Acris, Hyla, Pseudacris, Gastrophryne, Rana, Ambys-*

toma, Eurycea, Plethodon, Typhlotriton, Terrapene, Cnemidophorus, Elaphe, Lampropeltis, Nerodia, Agkistrodon, Crotalus.

117 **Black, J. H.** 1971. The formation of tadpole holes. Herpetol. Rev. 3(1):7.—An abstract. Behavior. *Bufo, Pseudacris.*

118 **Black, J. H.** 1973. Checklist of cave life of Oklahoma. I. Amphibia. Proc. Okla. Acad. Sci. 53:33-37.—Distribution, Natural History. *Bufo, Acris, Hyla, Pseudacris, Gastrophryne, Rana, Ambystoma, Eurycea, Plethodon, Typhlotriton.*

119 **Black, J. H.** 1973. Ethoecology of *Scaphiopus* (Pelobatidae) larvae in temporary pools in central and southwestern Oklahoma. Ph.D. Diss., Univ. Okla., Norman. 221 pp.—Behavior, Bibliographies, Development, Ecology. *Scaphiopus.*

120 **Black, J. H.** 1974. Larval spadefoot survival. J. Herpetol. 8(4):371-373.—Development, Natural History. *Scaphiopus.*

121 **Black, J. H.** 1974. Notes on *Plethodon ouachitae* in Oklahoma. Proc. Okla. Acad. Sci. 54:88-89.—Distribution, Natural History. *Plethodon.*

122 **Black, J. H.** 1974. Tadpole nests in Oklahoma. Okla. Geol. Notes 34(3):105.—Natural History. *Bufo, Pseudacris.*

123 **Black, J. H.** 1975. The formation of "tadpole nests" by anuran larvae. Herpetologica 31(1):76-79.—Natural History. *Bufo, Pseudacris.*

124 **Black, J. H.** 1975. Geographic distribution: *Deirochelys reticularia miaria.* Herpetol. Rev. 6(2):44.—Distribution. *Deirochelys.*

125 **Black, J. H.** 1975. Geographic distribution: *Kinosternon subrubrum hippocrepis.* Herpetol. Rev. 6(2):44.—Distribution. *Kinosternon.*

126 **Black, J. H.** 1975. Geographic distribution: *Sternotherus carinatus.* Herpetol. Rev. 6(2):44.—Distribution. *Sternotherus.*

127 **Black, J. H.** 1976. Escape behavior of the southern Plains skink. Herpetol. Rev. 7(3):111.—Behavior. *Eumeces.*

128 **Black, J. H.** 1976. Lined snake as food for birds. Bull. Okla. Ornithol. Soc. 9(3):17-18.—Natural History. *Tropidoclonion.*

129 **Black, J. H.** 1976. Oklahoma cave amphibians protected by state law. Bull. Okla. Herp. Soc. 1(4):67-68.—Management. *Eurycea, Typhlotriton.*

130 **Black, J. H.** 1976. Oklahoma leopard frogs. Bull. Okla. Herp. Soc. 1(1):6-10.—Distribution. *Rana.*

131 **Black, J. H.** 1976. Polydactylous Rocky Mountain toad. Bull. Okla. Herp. Soc. 1(3):41.—Morphology. *Bufo.*

132 **Black, J. H.** 1977. Herpetol. Rev. 8(4):132.—Photograph of bullfrog at Oklahoma City Zoo. Popular. *Rana.*

133 **Black, J. H.** 1977. Endangered and threatened amphibians and reptiles of Oklahoma. O. H. S. committee report. Bull. Okla. Herp. Soc. 2(2-3):42-50.—Management. *Bufo, Hyla, Scaphiopus, Rana, Ambystoma, Amphiuma, Eurycea, Hemidactylium, Plethodon, Typhlotriton, Siren, Chrysemys, Graptemys, Phrynosoma, Sceloporus, Uta, Cnemidophorus, Cemophora, Farancia, Hypsiglena, Regina, Storeria, Thamnophis, Virginia, Alligator.*

134 **Black, J. H.** 1977. Juvenile green anole in Pottawatomie Co., Oklahoma. Bull. Okla. Herp. Soc. 2(2-3):51.—Distribution. *Anolis.*

135 **Black, J. H.** 1978. History of Oklahoma Herpetology. IV. A. I. Ortenburger. 1926. Exploring Oklahoma's Unknown, Outdoor Oklahoma 2(10):6-7, 30-31. Bull. Okla. Herp. Soc. 3(2):35-40.—Reprint of a rare publication different from current Outdoor Oklahoma. General Reference, Historical.

136 **Black, J. H.** 1980. Acting champion of the Oklahoma snake world. Bull. Okla. Herp. Soc. 5(2):27-28.—Behavior, Lore, Natural History. *Heterodon.*

137 **Black, J. H.** 1980. Amphibians of Oklahoma-A checklist. Bull. Okla. Herp. Soc. 4(4):78-80.—Distribution. All Amphibia.

138 **Black, J. H.** 1980. "Pet frog." Bull. Okla. Herp. Soc. 5(2):29-30.—Popular. *Hyla.*

139 **Black, J. H.** 1981. Oklahoma rattlesnake hunts-1981. Bull. Okla. Herp. Soc. 6(2):39-43.—Management. *Crotalus.*

140 **Black, J. H.** 1981. Softshell turtles-candidates for Aquaticus. Zoosounds 17(5):4-5.—Natural History. *Trionyx.*

141 **Black, J. H.** 1982. An annotated bibliography to articles, notes and photographs on reptiles and amphibians appearing in Oklahoma Game and Fish News, Oklahoma Wildlife, and Outdoor Oklahoma. Volume 1 through Volume 37, 1945 through 1981. Okla. Herp. Soc. Spec. Publ. (2):1-53.—Bibliographies, General Reference.

142 **Black, J. H.** 1982. Softshell turtles candidate for aquaticus. Bull. Okla. Herp. Soc. 7(1):25-26.—Reprinted from Black (1981, Zoosounds). Natural History. *Trionyx.*

143 **Black, J. H.** 1983. Northern watersnakes eaten by channel catfish. Bull. Okla. Herp. Soc. 8(2):55-56.—Natural History. *Nerodia.*

144 **Black, J. H.** 1983. Red-tailed hawk captures prairie kingsnake. Bull. Okla. Herp. Soc. 8(2):63-65.—Natural History. *Lampropeltis.*

145 **Black, J. H.** 1985. Amphibians and reptiles associated with temporary pools near Tipton, Tillman County, Oklahoma. Bull. Okla. Herp. Soc. 10(1-2):5-6.—Distribution, Natural History. *Bufo, Acris, Pseudacris, Gastrophryne, Scaphiopus, Rana, Terrapene, Kinosternon, Phrynosoma, Cnemidophorus, Lampropeltis, Masticophis, Pituophis, Regina, Rhinocheilus.*

146 **Black, J. H.** 1985. Frog fever. Bull. Okla. Herp. Soc. 10(1-2):1-2.—Popular account of experience with Strecker's chorus frogs. Natural History, Popular. *Pseudacris.*

147 **Black, J. H.** 1985. The lined snake as food for the great horned owl. Bull. Okla. Ornithol. Soc. 18(2):14-15.—Natural History. *Tropidoclonion.*

148 **Black, J. H., and A. H. Black.** 1978. Brood of Graham's water snakes from Pottawatomie County, Oklahoma. Bull. Okla. Herp. Soc. 3(4): 69.—Distribution, Natural History. *Nerodia.*

149 **Black, J. H., P. E. Black, and A. H. Black.** 1983. County records for reptiles in central Oklahoma. Bull. Okla. Herp. Soc. 8(4):101-103.—Distribution, Natural History. *Bufo, Sternotherus, Ophisaurus, Crotaphytus, Cnemidophorus, Heterodon, Storeria.*

150 **Black, J. H., M. Butcher, F. A. Barkley, P. E. Black, A. H. Black, and D. Loving.** 1980. New county records of some Oklahoma reptiles. Bull. Okla. Herp. Soc. 5(1):7-8.—Distribution. *Pseudemys, Terrapene, Sceloporus, Eumeces, Scincella, Cnemidophorus, Diadophis, Elaphe, Hypsiglena, Pituophis, Sonora, Thamnophis, Virginia.*

151 **Black, J. H., M. Butcher, and E. M. Palmer.** 1979. Geographic distribution: *Eumeces anthracinus pluvialis.* Herpetol. Rev. 10(3):102.— Distribution. *Eumeces.*

152 **Black, J. H., and C. C. Carpenter.** 1969. Arthur Norris Bragg. Copeia 1969(2):419-420.—An obituary. General Reference, Historical.

153 **Black, J. H., T. Hunkapiller, and D. Dawson.** 1976. Winter activity in Oklahoma frogs. Bull. Okla. Herp. Soc. 1(2):22.—Natural History. *Acris, Rana.*

154 **Black, J. H., E. Latham, J. Pigg, and R. L. Lardie.** 1986. Distribution records and notes on *Trionyx* in Oklahoma. Bull. Okla. Herp. Soc. 11(1-4):42.—Distribution. *Trionyx.*

155 **Black, J. H., and D. W. Lollis.** 1979. Geographic distribution: *Virginia striatula.* Herpetol. Rev. 10(3):103.—Distribution. *Virginia.*

156 **Black, J. H., and W. L. Puckette.** 1974. The Quachita salamander in Oklahoma caves. Okla. Underground 6(2):42-44.—Natural History. *Plethodon.*

157 **Black, J. H., and S. M. Secor.** 1983. Three-toed box turtle in Haskell County, Oklahoma. Bull. Okla. Herp. Soc. 8(4):99-100.—Distribution. *Terrapene.*

158 **Black, J. H., and S. L. Shannon.** 1978. Five-lined skink from Okfuskee County, Oklahoma. Bull. Okla. Herp. Soc. 3(4):73.—Distribution. *Eumeces.*

159 **Black, J. H., and S. L. Shannon.** 1978. Geographic distribution: *Trionyx spiniferus.* Herpetol. Rev. 9(4):142.—Distribution. *Trionyx.*

160 **Black, J. H., and J. Short.** 1976. Geographic distribution: *Diadophis punctatus arnyi.* Herpetol. Rev. 7(3):123.—Distribution. *Diadophis.*

Black, J. H., *see* 94, 361, 391, 664, 674, 945, 975, 1031

161 **Black, P. E., A. H. Black, and G. Whitehead.** 1982. Oklahoma County record for the western spiny softshell. Bull. Okla. Herp. Soc. 7(1):3.— Distribution. *Trionyx.*

Black, P. E., *see* 149, 150

162 **Blair, A. P.** 1939. Records of the salamander *Typhlotriton.* Copeia 1939(2):109.—Distribution. *Typhlotriton.*

163 **Blair, A. P.** 1941. *Ambystoma talpoideum* in Oklahoma. Copeia 1941(3):184.—Distribution. *Ambystoma.*

164 **Blair, A. P.** 1941. Variation, isolation mechanisms and hybridization in certain toads. Genetics 26(July):398-417.—Behavior, Development, Ecology, Genetics, Morphology, Physiology, Taxonomy. *Bufo.*

165 **Blair, A. P.** 1943. Geographical variation of ventral markings of toads. Am. Midl. Nat. 29(2):615-620.—Morphology. *Bufo.*

166 **Blair, A. P.** 1943. Population structure in toads. Am. Nat. 77(733): 563-568.—Behavior, Ecology. *Bufo.*

167 **Blair, A. P.** 1950. The alligator in Oklahoma. Copeia 1950(1):57.—Distribution. *Alligator.*

168 **Blair, A. P.** 1950. Notes on Oklahoma microhylid frogs. Copeia 1950(2):152.—Distribution, Genetics, Natural History. *Gastrophryne.*

169 **Blair, A. P.** 1950. Notes on two anguid lizards. Copeia 1950(1):57.—Behavior. *Ophisaurus.*

170 **Blair, A. P.** 1950. Skittering locomotion in *Acris crepitans.* Copeia 1950(3):237.—Natural History. *Acris.*

171 **Blair, A. P.** 1950. Some cold-blooded vertebrates of the Oklahoma panhandle. Copeia 1950(3):234.—Distribution. *Bufo, Rana, Ambystoma, Kinosternon, Holbrookia, Phrynosoma, Sceloporus, Eumeces, Cnemidophorus, Tropidoclonion, Crotalus.*

172 **Blair, A. P.** 1951. Note on Oklahoma salamanders. Copeia 1951(2): 178.—Distribution, Natural History. *Typhlotriton, Siren, Notophthalmus.*

173 **Blair, A. P.** 1951. Winter activity in Oklahoma frogs. Copeia 1951(2):178.—Natural History. *Acris, Pseudacris, Rana.*

174 **Blair, A. P.** 1952. Notes on amphibians from Oklahoma and North Dakota. Copeia 1952(2):114-115.—Distribution. *Hyla, Gastrophryne, Ambystoma, Eurycea, Plethodon, Typhlotriton, Notophthalmus.*

175 **Blair, A. P.** 1957. A comparison of living *Plethodon ouachitae* and *Plethodon caddoensis.* Copeia 1957(1):47-48.—Morphology, Taxonomy. *Plethodon.*

176 **Blair, A. P.** 1961. Notes on *Ophisaurus attenuatus attenuatus* (Anguidae). Southwest. Nat. 6(3-4):201.—Distribution. *Ophisaurus.*

177 **Blair, A. P.** 1963. Notes on anuran behavior, especially *Rana catesbeiana.* Herpetologica 19(2):151.—Behavior. *Rana.*

178 **Blair, A. P.** 1967. Bullfrog attempts to catch dove. Southwest. Nat. 12(2):201-202.—Natural History. *Rana.*

179 **Blair, A. P.** 1967. *Plethodon ouachitae.* Cat. Am. Amphib. Reptiles 40.1-40.2.—Range map and literature review. Bibliographies, Distribution. *Plethodon.*

180 **Blair, A. P., and H. D. Chase.** 1935. Amphibia found in a survey of Tulsa County and vicinity. Proc. Okla. Acad. Sci. 16:35-36.—Distribution. *Bufo, Acris, Hyla, Pseudacris, Gastrophryne, Rana, Ambystoma, Eurycea, Necturus, Notophthalmus.*

181 **Blair, A. P., and H. E. Laughlin.** 1955. Range extensions of *Gastrophryne carolinensis.* Copeia 1955(4):311.—Distribution. *Gastrophryne.*

182 **Blair, A. P., and H. L. Lindsay, Jr.** 1961. *Hyla avivoca* (Hylidae) in Oklahoma. Southwest. Nat. 6(3-4):202.—Distribution, Natural History. *Hyla, Rana, Ambystoma, Plethodon, Notophthalmus, Anolis, Farancia, Agkistrodon.*

183 **Blair, A. P., and H. L. Lindsay, Jr.** 1965. Color pattern variation and distribution of two large *Plethodon* salamanders endemic to the Ouachita Mountains of Oklahoma and Arkansas. Copeia 1965(3): 331-335.—Distribution, Morphology, Taxonomy. *Plethodon.*

Blair, A. P., *see* 197, 498

184 **Blair, W. F.** 1936. A note on the ecology of *Microhyla olivacea.* Copeia 1936(2):115.—Ecology. *Gastrophryne.*

185 **Blair, W. F.** 1955. Differentiation of mating call in spadefoots, genus *Scaphiopus.* Tex. J. Sci. 7(2):183-188.—Behavior, Evolution. *Gastrophryne.*

186 **Blair, W. F.** 1955. Mating call and stage of speciation in the *Microhyla olivacea-M. carolinensis* complex. Evolution 9(4):469-480.—Behavior, Evolution. *Gastrophryne.*

187 **Blair, W. F.** 1955. Size difference as a possible isolation mechanism in *Microhyla.* Am. Nat. 99(848):297-301.—Behavior, Evolution, Morphology. *Gastrophryne.*

188 **Blair, W. F.** 1956. Call difference as an isolation mechanism in southwestern toads (Genus *Bufo*). Tex. J. Sci. 8(1):87-106.—Behavior, Evolution. *Bufo.*

189 **Blair, W. F.** 1956. The mating calls of hybrid toads. Tex. J. Sci. 8(3): 350-355.—Oklahoma toads used in crosses. Genetics. *Bufo.*

190 **Blair, W. F.** 1958. Distributional patterns of vertebrates in the southern United States in relation to past and present environments. Pages 433-468 *in* C. L. Hubbs (Ed.). Zoogeography. Am. Assoc. Adv. Sci. (51):509.—Distribution, Paleontology. *Bufo, Pseudacris, Gastrophryne, Eurycea, Plethodon, Eumeces.*

191 **Blair, W. F.** 1958. Mating call in the speciation of anuran amphibians. Am. Nat. 92(862):27-51.—Distribution, Evolution. *Bufo, Acris, Hyla, Pseudacris, Gastrophryne.*

192 **Blair, W. F.** 1961. Further evidence bearing on intergroup and intragroup genetic compatibility in toads (Genus *Bufo*). Tex. J. Sci. 13(2): 163-175.—Evolution, Genetics. *Bufo.*

193 **Blair, W. F.** 1962. Non-morphological data in anuran classification. Syst. Zool. 11(2):72-84.—Behavior, Distribution, Taxonomy. *Bufo.*

194 **Blair, W. F.** 1965. Amphibian speciation. Pages 543-556 *in* H. E. Wright Jr. and D. G. Frey (Eds.). The quaternary of the United States. Princton Univ. Press, Princeton. 922 pp.—Distribution, Evolution, Paleontology, *Bufo, Scaphiopus, Plethodon.*

195 **Blair, W. F.** 1972. Evidence from hybridization. Pages 196-232 *in* W. F. Blair (Ed.). Evolution in the genus *Bufo.* Univ. Tex. Press, Austin. 459 pp.—Evolution. *Bufo.*

196 **Blair, W. F.** 1972. Evolution in the genus *Bufo.* Univ. Tex. Press, Austin. 459 pp.—Behavior, Bibliographies, Morphology, Techniques. *Bufo.*

197 **Blair, W. F., and A. P. Blair.** 1941. Food habits of the collared lizard in northeastern Oklahoma. Am. Midl. Nat. 26(1):230-232.—Natural History. *Crotaphytus.*

198 **Blanchard, F. N.** 1920. A synopsis of the king snakes: Genus *Lampropeltis* Fitzinger. Occas. Pap. Mus. Zool. Univ. Mich. (87):1-7.—A key for the genus. Distribution, Taxonomy. *Lampropeltis.*

199 **Blanchard, F. N.** 1921. A revision of the king snakes: Genus *Lampropeltis.* Bull. U. S. Natl. Mus. (114):1-260.—Bibliographies, Distribution, Historical, Taxonomy. *Lampropeltis.*

200 **Blanchard, F. N.** 1923. The snakes of the genus *Virginia.* Pap. Mich. Acad. Sci. Arts Lett. (3):343-365.—Behavior, Distribution, Ecology, Morphology, Natural History, Taxonomy, *Virginia.*

201 **Blanchard, F. N.** 1924. The status of *Amphiardis inornatus* (Garman). Copeia 1924(134):83-85.—Taxonomy. *Virginia.*

202 **Blanchard, F. N.** 1932. A clutch of eggs of the speckled king snake, *Lampropeltis getulus holbrooki* (Stejneger). Copeia 1932(2):98.—Article includes time until hatching and length of hatchling speckled king snakes. Female was from Tulsa. Natural History. *Lampropeltis.*

203 **Blanchard, F. N.** 1938. Snakes of the genus *Tantilla* in the United States. Zool. Ser. Field Mus. Nat. Hist. 20(28):369-376.—Distribution, Taxonomy. *Tantilla.*

204 **Blanchard, F. N.** 1942. The ring-neck snakes, genus *Diadophis*. Bull. Chic. Acad. Sci. 7(1):1-144.—Distribution, Ecology, Natural History. *Diadophis*.

205 **Blanchard, F. N., and E. R. Force.** 1930. The age of attainment of sexual maturity in the lined snake, *Tropidoclonion lineatum* (Hallowell). Bull. Antivenin Inst. Am. 3(4):96-98.—Natural History. *Tropidoclonion*.

206 **Blaney, R. M.** 1973. *Lampropeltis*. Cat. Am. Amphib. Reptiles 150.1-150.2.—Range map and literature review. Bibliographies, Distribution. *Lampropeltis*.

207 **Blaney, R. M.** 1973. Systematics of the common kingsnake *Lampropeltis getulus* (Linnaeus). Tulane Stud. Zool. Bot. 19(3-4):47-104.—Distribution, Taxonomy. *Lampropeltis*.

208 **Blaney, R. M.** 1979. *Lampropeltis calligaster*. Cat. Am. Amphib. Reptiles 229.1-229.2.—Range map and literature review. Bibliographies, Distribution. *Lampropeltis*.

209 **Bleakney, S.** 1958. Postglacial dispersal of the turtle *Chrysemys picta*. Herpetologica 14(2):101-104.—Distribution. *Chrysemys*.

210 **Blem, C. R.** 1981. *Heterodon platyrhinos*. Cat. Am. Amphib. Reptiles 282.1-282.2.—Range map and literature review. Bibliographies, Distribution. *Heterodon*.

211 **Blem, C. R.** 1987. Range portrayal and reality: "Bias" or scaling problem? Herpetol. Rev. 18(1):9.—Included range of hog-nosed snakes in Oklahoma. Distribution, Techniques. *Heterodon*.

212 **Bobalik, S. J., and J. R. Galm.** 1978. Appendix C: Faunal remains from the Scott (34Lf-11) and Wann (34Lf-27) sites, Oklahoma. Pages 299-332 *in* J. R. Galm and P. Flynn (Eds.). The cultural sequences at the Scott (34Lf-11) and Wann (34Lf-27) sites and prehistory of the Wister Valley. Univ. Okla. Archeol. Res. Manage. Cent. Res. Ser. (3):1-350.—Archeology. *Chelydra, Graptemys, Trachemys, Terrapene, Trionyx*.

213 **Bolt, J. R.** 1948. Armor of dissophorids (Amphibia, Labyrinthodontia): An examination of its taxonomic use and report on a new occurrence. J. Paleontol. 48(1):135-142.—Armor scutes from the Lower Permian, Fort Sill. General Reference, Paleontology.

214 **Bolt, J. R.** 1948. Evolution and functional interpretation of some suture patterns in paleozoic labyrinthodont amphibians and other lower

tetrapods. J. Paleontol. 48(3):434-458.—Some specimens from Fort Sill, Oklahoma. General Reference, Paleontology.

215 **Bolt, J. R.** 1968. The osteology and relationships of *Doleserpeton annectens,* a new rhachitomons amphibian from The Lower Permian of Oklahoma. Ph.D. Diss., Univ. Chicago, 90 pp.—General Reference, Paleontology.

216 **Bolt, J. R.** 1969. Lissamphibian origins: Possible protoliss amphibian from the Lower Permian of Oklahoma. Science 166(3907):888-891.—General Reference, Paleontology.

217 **Bolt, J. R.** 1974. A trematopsid skull from the Lower Permian, and analysis of some characters of the dissorophoid (Amphibia: Labyrinthodontia) otic notch. Fieldiana Geol. 30(3):67-79.—Specimen from Lower Permian of Fort Sill. General Reference, Morphology, Paleontology.

218 **Bolt, J. R.** 1975. Lower tetropods of 1/4 billion years ago. Bull. Field Mus. Nat. Hist. 46(10):8-11.—General Reference, Paleontology.

219 **Bolt, J. R.** 1977. *Cacops* (Amphibia: Labyrinthodontia) from the Fort Sill locality, Lower Permian of Oklahoma. Fieldiana Geol. 37(3):67-73.—General Reference, Paleontology.

220 **Bolt, J. R.** 1977. Dissorophoid relationships and ontogeny, and the origin of the Lissamphibia. J. Paleontol. 51(2):235-249.—All specimens from Lower Permian deposits in Oklahoma and Texas. General Reference, Paleontology.

221 **Bolt, J. R.** 1980. New tetrapods with bicuspid teeth from the Fort Sill locality (Lower Permian, Oklahoma). Neues. Jb. Geol. Paleontol. Mh. 1980(8):449-459.—Included Oklahoma reptiles. General Reference, Paleontology.

222 **Bolt, J. R., and R. DeMar.** 1975. An explanatory model of the evolution of multiple rows of teeth in *Captorhinus aguti.* J. Paleontol. 49(5): 814-832.—Fossil amphibians from Fort Sill fissure fill deposits, Lower Permian, Oklahoma. General Reference, Paleontology.

223 **Bolt, J. R., and R. J. Wasserzug.** 1975. Functional morphology of the skull of *Lysorophus*: A snake-like Paleozoic amphibian (Lepospondyli). Paleobiology 1(3):320-332.—General Reference, Morphology, Paleontology.

Bolt, J. R., *see* 559

224 **Bonn, E. W., and H. W. McCarley.** 1953. The amphibians and reptiles of the Lake Texoma area. Tex. J. Sci. 5(4):465-471.—Distribution. All Anura, *Siren, Notophthalmus, Chelydra, Graptemys, Pseudemys, Trachemys, Terrapene, Kinosternon, Trionyx, Ophisaurus, Phrynosoma, Sceloporus, Eumeces, Scincella, Cnemidophorus, Coluber, Diadophis, Elaphe, Heterodon, Lampropeltis, Masticophis, Nerodia, Opheodrys, Pituophis, Storeria, Thamnophis, Tropidoclonion, Virginia, Agkistrodon, Crotalus, Sistrurus.*

225 **Bontrager, S. K.** 1977. Autecology of *Crotaphytus collaris.* M.S. Thesis, Okla. State Univ., Stillwater. 151 pp.—Ecology. *Crotaphytus.*

226 **Bouchard, J. L.** 1953. An ecological and taxonomic study of helminth parasites from Oklahoma amphibians. Ph.D. Diss., Univ. Okla., Norman. 138 pp.—Distribution, Parasitology. All Anura, *Ambystoma, Desmognathus, Eurycea, Plethodon, Notophthalmus.*

Bowman, M. C., *see* 661

227 **Boyer, D. R.** 1953. Variation in a population of *Crotalus atrox,* Baird and Girard in Comanche County, Oklahoma. M.S. Thesis, Univ. Okla., Norman. 73 pp.—Morphology. *Crotalus.*

228 **Boyer, D. R.** 1957. Sexual dimorphism in a population of the western diamond-backed rattlesnake. Herpetologica 13(3):213-217.—The 214 rattlesnakes in this study were collected in the Wichita Mountain Wildlife Refuge. Morphology. *Crotalus.*

229 **Brackin, M. F.** 1975. The relation of status of behavior and anatomy in dominance hierarchies of six-lined racerunner lizards. Ph.D. Diss., Univ. Okla., Norman. 41 pp.—Behavior, Morphology, Physiology. *Cnemidophorus.*

230 **Brackin, M. F.** 1978. The relation of rank to physiological state in *Cnemidophorus sexlineatus* dominance hierarchies. Herpetologica 34(2):185-191.—Behavior, Physiology. *Cnemidophorus.*

231 **Brackin, M. F.** 1979. The seasonal reproductive, fat body and adrenal cycles of male six-lined racerunners (*Cnemidophorus sexlineatus*) in central Oklahoma. Herpetologica 35(3):216-222.—Morphology, Physiology. *Cnemidophorus.*

232 **Bragg, A. N.** 1936. The ecological distribution of some North American Anura. Am. Nat. 70(730):459-466.—Distribution, Ecology. *Acris, Rana.*

233 **Bragg, A. N.** 1936. Notes on the breeding habits, eggs, and embryos of *Bufo cognatus* with a description of the tadpole. Copeia 1936(1): 14-20.—Behavior, Development, Morphology. *Bufo.*

234 **Bragg, A. N.** 1937. A note on the metamorphosis of the tadpoles of *Bufo cognatus.* Copeia 1937(4):227-228.—Development. *Bufo.*

235 **Bragg, A. N.** 1937. Observations on *Bufo cognatus* with special reference to the breeding habits and eggs. Am. Midl. Nat. 18(2):273-284.—Behavior, Development, Morphology, Natural History. *Bufo.*

236 **Bragg, A. N.** 1937. A study of the mitotic cycle and mitotic distribution in the early embryo of *Bufo cognatus* Say with special reference to factors in the organization of the embryo. Ph.D. Diss., Univ. Okla., Norman. 125 pp.—Development, Morphology. *Bufo.*

237 **Bragg, A. N.** 1938. Observations on the natural history of *Bufo cognatus* Say. Proc. Okla. Acad. Sci. 19:41-42.—Behavior, Development, Ecology, Morphology, Natural History. *Bufo, Scaphiopus.*

238 **Bragg, A. N.** 1938. The organization of the early embryo of *Bufo cognatus* as revealed especially by the mitotic index. Z. Zellforsch. Mikrosk. Anat. 28(2):154-178.—Development, Techniques. *Bufo.*

239 **Bragg, A. N.** 1939. Dwarf *Bufo americanus americanus* from Cleveland County, Oklahoma. Proc. Okla. Acad. Sci. 20:75-76.—Morphology, Natural History, Taxonomy. *Bufo.*

240 **Bragg, A. N.** 1939. Observations on the ecology and natural history of Anura. V. The process of hatching in several species. Proc. Okla. Acad. Sci. 20:71-74.—Development. *Bufo, Hyla, Gastrophryne, Rana.*

241 **Bragg, A. N.** 1939. Observations upon amphibian deutoplasm and its relation to embryonic and early larval development. Biol. Bull. 72(2): 268-284.—Development. *Bufo, Scaphiopus, Rana.*

242 **Bragg, A. N.** 1939. Possible hybridization between *Bufo cognatus* and *B. w. woodhousii.* Copeia 1939(3):173.—Development. *Bufo.*

243 **Bragg, A. N.** 1939. Some cytological phenomena in early embryos of *Bufo cognatus* Say. Trans. Am. Microsc. Soc. 58(3):357-370.—Development. *Bufo.*

244 **Bragg, A. N.** 1940. Key to the toads (*Bufo*) of Oklahoma. Proc. Okla. Acad. Sci. 21:17-18.—Morphology, Taxonomy. *Bufo.*

245 **Bragg, A. N.** 1940. Observations on the ecology and natural history of Anura. III. The ecological distribution of Anura of Cleveland County, Oklahoma, including notes on the habits of several species.

Am. Midl. Nat. 24(2):322-335.—Behavior, Ecology, Natural History. All Anura.

246 **Bragg, A. N.** 1940. Observations on the ecology and natural history of Anura. II. Habits, habitat, and breeding of *Bufo woodhousii woodhousii* (Girard) in Oklahoma. Am. Midl. Nat. 24(2):306-321.—Behavior, Ecology, Natural History. *Bufo, Pseudacris, Scaphiopus, Rana.*

247 **Bragg, A. N.** 1940. Observations on the ecology and natural history of Anura. VI. The ecological importance of the study of the habits of animals as illustrated by toads. Wasmann Club Coll. 4(1):6-16.—Behavior, Ecology, Natural History. *Bufo, Scaphiopus.*

248 **Bragg, A. N.** 1941. A corrected list of the Amphibia known in central Oklahoma. Proc. Okla. Acad. Sci. 22:16-17.—Distribution. All Anura, *Ambystoma.*

249 **Bragg, A. N.** 1941. A key to the frogs of the genus *Rana* in Oklahoma. Proc. Okla. Acad. Sci. 22:18.—Morphology, Taxonomy. *Rana.*

250 **Bragg, A. N.** 1941. New records of frogs and toads for Oklahoma. Copeia 1941(1):51-52.—Distribution. All Anura.

251 **Bragg, A. N.** 1941. Observations on the ecology and natural history of Anura. VIII. Some factors in the initiation of breeding behavior. Turtox News 19(1):10-12.—Natural History, Physiology. *Bufo, Hyla, Pseudacris, Gastrophryne, Scaphiopus, Rana.*

252 **Bragg, A. N.** 1941. Observations on the ecology and natural history of Anura. XI. The invasion of the Canadian River Flood Plain by two prairie species. Proc. Okla. Acad. Sci. 22:73-74.—Ecology, Natural History. *Bufo, Pseudacris.*

253 **Bragg, A. N.** 1941. Tadpoles of *Scaphiopus bombifrons* and *Scaphiopus hammondii.* Wasmann Club Coll. 4(3):92-94.—Development, Morphology. *Scaphiopus.*

254 **Bragg, A. N.** 1942. Common names for frogs and toads in Oklahoma. Proc. Okla. Acad. Sci. 23:39-40.—Natural History. All Anura.

255 **Bragg, A. N.** 1942. The ecological distribution of the hylid and microhylid frogs in Oklahoma. Bull. Ecol. Soc. Am. 23(4):69.—An abstract. Distribution, Ecology. *Acris, Hyla, Pseudacris, Gastrophryne.*

256 **Bragg, A. N.** 1942. On the economic value of Oklahoma toads. Proc. Okla. Acad. Sci. 23:37-39.—Discussed the value of toads in controlling insects. Economics, Natural History. *Bufo.*

257 **Bragg, A. N.** 1942. Further field notes on the initiation of breeding behavior of Anura. Turtox News 20(1):12-13.—Behavior. *Bufo, Pseudacris, Gastrophryne, Scaphiopus.*

258 **Bragg, A. N.** 1942. Growth rates and age at sexual maturity of young toads (*Bufo cognatus*) under natural conditions in central Oklahoma. Bull. Ecol. Soc. Am. 23(4):69.—An abstract. Natural History. *Bufo.*

259 **Bragg, A. N.** 1942. The habits of spadefoot toads in relation to their phylogeny. Anat. Rec. 84(4):507.—An abstract. Natural History. *Scaphiopus.*

260 **Bragg, A. N.** 1942. Key to spadefoot toads (*Scaphiopus*) in Oklahoma. Turtox News 20(11):154.—Taxonomy. *Scaphiopus.*

261 **Bragg, A. N.** 1942. Life history of Hurter's Spadefoot. Anat. Rec. 84(4):506.—An abstract. Natural History. *Scaphiopus.*

262 **Bragg, A. N.** 1942. Observations on the ecology and natural history of Anura. X. The breeding habits of *Pseudacris streckeri* Wright and Wright in Oklahoma including a description of the eggs and tadpoles. Wasmann J. Biol. 5(2):47-62.—Behavior, Development, Ecology, Morphology, Natural History. *Pseudacris.*

263 **Bragg, A. N.** 1942. On toad and frog abundance after heavy rainfall. Science 95(2460):194-195.—Behavior, Ecology, Natural History. All Anura.

264 **Bragg, A. N.** 1943. Egg laying in leopard frogs. Proc. Okla. Acad. Sci. 24:13-14.—Behavior, Natural History. *Rana.*

265 **Bragg, A. N.** 1943. Habits in the ecological distribution of Oklahoma Salientia. Anat. Rec. 87(4):435-436.—An abstract. Ecology. *Scaphiopus.*

266 **Bragg, A. N.** 1943. Observations on the ecology and natural history of Anura. XV. The hylids and microhylids in Oklahoma. Great Basin Nat. 4(3-4):62-80.—Ecology, Natural History, Taxonomy. *Acris, Hyla, Pseudacris, Gastrophryne.*

267 **Bragg, A. N.** 1943. Observations on the ecology and natural history of Anura. XVI. Life-history of *Pseudacris clarkii* (Baird) in Oklahoma. Wasmann Club Coll. 5(4):129-140.—Behavior, Development, Ecology, Morphology, Natural History. *Bufo, Hyla, Pseudacris, Gastrophryne, Scaphiopus.*

268 **Bragg, A. N.** 1944. Breeding habits, eggs, and tadpoles of *Scaphiopus hurterii.* Copeia 1944(4):230-241.—Behavior, Development, Distribution, Ecology, Morphology, Natural History, *Scaphiopus.*

269 **Bragg, A. N.** 1944. The spadefoot toads in Oklahoma with a summary of our knowledge of the group. Am. Nat. 78(779):517-533.—Behavior, Distribution, Ecology, Morphology, Natural History, Taxonomy. *Scaphiopus.*

270 **Bragg, A. N.** 1945. Aggregational phenomena in *Scaphiopus hurterii* tadpoles. Proc. Okla. Acad. Sci. 26:19.—Natural History. *Scaphiopus.*

271 **Bragg, A. N.** 1945. Breeding and tadpole behavior in *Scaphiopus hurterii* near Norman, Oklahoma, spring, 1945. Wasmann Club Coll. 6(3-4):69-78.—Behavior, Natural History. *Bufo, Pseudacris, Scaphiopus.*

272 **Bragg, A. N.** 1945. Notes on the psychology of frogs and toads. J. Gen. Psychol. 32(1):27-37.—Natural History. All Anura.

273 **Bragg, A. N.** 1945. Puzzles in the life history of *Scaphiopus hurterii* Strecker. Proc. Okla. Acad. Sci. 25:27.—Natural History. *Scaphiopus.*

274 **Bragg, A. N.** 1945. The spadefoot toads in Oklahoma with a summary of our knowledge of the group II. Am. Nat. 79(780):52-72.—Behavior, Development, Ecology, Natural History, Taxonomy. *Bufo, Hyla, Pseudacris, Gastrophryne, Scaphiopus, Rana.*

275 **Bragg, A. N.** 1946. Aggregation in tadpoles of spadefoot toads. Anat. Rec. 94(4):351-352.—An abstract. Behavior. *Scaphiopus.*

276 **Bragg, A. N.** 1946. Aggregation with cannibalism in tadpoles of *Scaphiopus bombifrons* with some general remarks on the probable evolutionary significance of such phenomena. Herpetologica 3(3):89-97.—Behavior, Evolution. *Scaphiopus.*

277 **Bragg, A. N.** 1946. Some new county records of Salientia and a correction in the list from Oklahoma. Proc. Okla. Acad. Sci. 26:16-18.—Distribution. *Bufo, Acris, Hyla, Pseudacris, Gastrophryne, Rana.*

278 **Bragg, A. N.** 1946. Some salientian adaptations. Anat. Rec. 96(4): 518.—An abstract. Ecology, Distribution, General Reference.

279 **Bragg, A. N.** 1946. Some salientian adaptations. Great Basin Nat. 7(1-4):11-15.—Distribution, Ecology, Evolution. *Bufo, Hyla, Pseudacris, Gastrophryne, Scaphiopus, Rana.*

280 **Bragg, A. N.** 1946. Tadpole behavior in pools and streams. Proc. Okla. Acad. Sci. 27:59-61.—Behavior, Morphology, Taxonomy. All Anura.

281 **Bragg, A. N.** 1947. Additional instances of social aggregation in tadpoles. Anat. Rec. 99(4):616.—An abstract. Behavior. *Scaphiopus.*

282 **Bragg, A. N.** 1947. The canebrake rattler in Oklahoma. Proc. Okla. Acad. Sci. 28:36-37.—Distribution, Morphology. *Crotalus.*

283 **Bragg, A. N.** 1947. Comment on mouthparts of tadpoles of *Bufo cognatus* and *Bufo compactilis.* Science 106(2747):166.—Morphology. *Bufo.*

284 **Bragg, A. N.** 1947. An observation showing the biological necessity for size selection of food objects by young toads. Proc. Okla. Acad. Sci. 28:35-36.—Natural History. *Bufo.*

285 **Bragg, A. N.** 1947. Observations on *Hyla versicolor* in Oklahoma. Proc. Okla. Acad. Sci. 28:31-35.—Morphology, Natural History, Taxonomy. *Hyla.*

286 **Bragg, A. N.** 1948. Additional instances of social aggregation in tadpoles. Wasmann Club Coll. 7(2):65-79.—Behavior. *Scaphiopus, Rana.*

287 **Bragg, A. N.** 1948. Food resources utilized by tadpoles of *Scaphiopus hurterii* in temporary pools. Anat. Rec. 101(4):706.—An abstract. Behavior, Natural History. *Scaphiopus.*

288 **Bragg, A. N.** 1948. Observations on the life history of *Pseudacris triseriata* (Wied) in Oklahoma. Wasmann Club Coll. 7(4):149-168.—Behavior, Development, Distribution, Morphology, Natural History. *Pseudacris.*

289 **Bragg, A. N.** 1948. A sexual difference in dermal secretion in spadefoot toads. Proc. Okla. Acad. Sci. 29:11-12.—Evolution. *Scaphiopus.*

290 **Bragg, A. N.** 1949. Field preservation of amphibian eggs and larvae. Turtox News 27(12):262-263.—Techniques. *Acris, Hyla, Pseudacris.*

291 **Bragg, A. N.** 1949. A note on the egg-masses of *Scaphiopus hurterii* Strecker. Proc. Okla. Acad. Sci. 30:18-19.—Natural History. *Scaphiopus.*

292 **Bragg, A. N.** 1949. Observations on the narrow-mouthed salamander. Proc. Okla. Acad. Sci. 30:21-24.—Behavior, Development, Distribution, Ecology, Natural History, *Bufo, Ambystoma.*

293 **Bragg, A. N.** 1949. Recording animal calls for teaching aids. Proc. Okla. Acad. Sci. 30:205-206.—Techniques. *Scaphiopus, Rana.*

294 **Bragg, A. N.** 1949. The status of the leopard frogs of Oklahoma. Wasmann Club Coll. 7(5):211-214.—Distribution, Taxonomy. *Rana.*

295 **Bragg, A. N.** 1950. Data on size range in adults of *Bufo w. woodhousii* Girard. Proc. Okla. Acad. Sci. 31:39-40.—Morphology, Natural History. *Bufo.*

296 **Bragg, A. N.** 1950. Frequency of sex calls in Salientia. Pages 117-125 *in* Researches on the Amphibia of Oklahoma. Univ. Okla. Press, Norman.—Behavior. *Bufo, Acris, Hyla, Pseudacris, Gastrophryne, Scaphiopus.*

297 **Bragg, A. N.** 1950. Further notes on range extensions of Salientia. Proc. Okla. Acad. Sci. 31:42-44.—Distribution, Natural History. *Bufo, Hyla, Pseudacris, Gastrophryne, Scaphiopus, Rana.*

298 **Bragg, A. N.** 1950. The identification of Salientia in Oklahoma. Pages 9-11 *in* Researches on the Amphibia of Oklahoma. Univ. Okla. Press, Norman.—Taxonomy. *Bufo, Acris, Hyla, Rana.*

299 **Bragg, A. N.** 1950. Keys to frogs and toads of Oklahoma. Pages 11-29 *in* Researches on the Amphibia of Oklahoma. Univ. Okla. Press, Norman.—Taxonomy. *Bufo, Acris, Pseudacris, Gastrophryne, Scaphiopus, Rana.*

300 **Bragg, A. N.** 1950. Mass movement at metamorphosis in the savannah spadefoot, *Scaphiopus hurterii* Strecker. Proc. Okla. Acad. Sci. 31:26-27.—Natural History. *Scaphiopus.*

301 **Bragg, A. N.** 1950. Observations on the ecology and natural history of Anura. XVII. Adaptations and distribution in accordance with habits in Oklahoma. Pages 59-100 *in* Researches on the Amphibia of Oklahoma. Univ. Okla. Press, Norman.—Behavior, Ecology, Natural History. All Anura.

302 **Bragg, A. N.** 1950. Observations on *Microhyla* (Salientia: Microhylidae). Wasmann J. Biol. 8(1):113-118.—Behavior, Development, Distribution, Taxonomy. *Bufo, Pseudacris, Gastrophryne, Scaphiopus.*

303 **Bragg, A. N.** 1950 Observations on *Scaphiopus,* 1949 (Salientia: Scaphiopodidae). Wasmann J. Biol. 8(2):221-228.—Discussion of social aggregations and cannibalism in tadpoles. Natural History. *Scaphiopus.*

304 **Bragg, A. N.** 1950. Salientian breeding dates in Oklahoma. Pages 34-38 *in* Researches on the Amphibia of Oklahoma. Univ. Okla. Press, Norman.—Behavior, Ecology. All Anura.

305 **Bragg, A. N.** 1950. Salientian range extensions in Oklahoma and a new state record. Pages 39-44 *in* Researches on the Amphibia of Oklahoma. Univ. Okla. Press, Norman.—Distribution, All Anura.

306 **Bragg, A. N.** 1950. Size range in adults of the toad *Bufo cognatus*. Copeia 1950(2):153-154.—Morphology. *Bufo*.

307 **Bragg, A. N.** 1950. Some adaptations of survival value in spadefoot toads. Pages 101-116 *in* Researches on the Amphibia of Oklahoma. Univ. Okla. Press, Norman.—Behavior, Ecology, Evolution. *Scaphiopus*.

308 **Bragg, A. N.** 1952. Amphibians of McCurtain County, Oklahoma. Wasmann J. Biol. 10(2):241-250.—Distribution. All Anura, *Ambystoma, Amphiuma, Desmognathus, Eurycea, Plethodon, Siren, Notophthalmus*.

309 **Bragg, A. N.** 1951. An apparatus for collecting small aquatic organisms. Turtox News 29(3):74-75.—General Reference, Techniques.

310 **Bragg, A. N.** 1952. Decline in toad populations in central Oklahoma. Proc. Okla. Acad. Sci. 33:70.—Natural History. *Bufo, Pseudacris, Scaphiopus*.

311 **Bragg, A. N.** 1952. A metamorphic aggregation in tadpoles of *Bufo*. Proc. Okla. Acad. Sci. 33:71.—Development, Natural History. *Bufo, Scaphiopus, Rana*.

312 **Bragg, A. N.** 1953. A study of *Rana areolata* in Oklahoma. Wasmann J. Biol. 11(3):273-318.—Behavior, Development, Distribution, Natural History. *Rana*.

313 **Bragg, A. N.** 1954. Aggregational behavior and feeding reactions in tadpoles of the savannah spadefoot. Herpetologica 10(2):97-102.—Behavior. *Scaphiopus*.

314 **Bragg, A. N.** 1954. *Bufo terrestris charlesmithi*, a new subspecies from Oklahoma. Wasmann J. Biol. 12(2):245-254.—Distribution, Natural History, Taxonomy. *Bufo*.

315 **Bragg, A. N.** 1954. A second season with *Rana areolata* in Oklahoma. Wasmann J. Biol. 12(3):337-343.—Behavior, Morphology, Natural History. *Rana*.

316 **Bragg, A. N.** 1955. The Amphibia of Cherokee County, Oklahoma. Herpetologica 11(1):25-30.—Distribution, Natural History. All Anura, *Ambystoma, Eurycea, Plethodon, Typhlotriton, Necturus, Notophthalmus.*

317 **Bragg, A. N.** 1955. The Amphibia of Greer County, Oklahoma. Great Basin Nat. 15(1-4):27-31.—Distribution. *Bufo, Acris, Pseudacris, Gastrophryne, Scaphiopus, Rana, Ambystoma.*

318 **Bragg, A. N.** 1955. Location of the types of *Bufo terrestris charlesmithi.* Herpetologica 11(2):151.—Taxonomy, *Bufo.*

319 **Bragg, A. N.** 1955. Supplementary notes on the young of *Rana areolata.* Herpetologica 11(2):87-88.—Development, Ecology, Natural History. *Pseudacris, Rana, Ambystoma.*

320 **Bragg, A. N.** 1955. The tadpole of *Bufo debilis debilis.* Herpetologica 11(3):211-212.—Morphology. *Bufo, Pseudacris.*

321 **Bragg, A. N.** 1955. Taxonomic and physiological factors in the embryonic development of certain toads. Copeia 1955(1):62.—Development, Distribution, Physiology, Taxonomy. *Bufo.*

322 **Bragg, A. N.** 1956. Dimorphism and cannibalism in tadpoles of *Scaphiopus bombifrons* (Amphibia; Salientia). Southwest. Nat. 1(3): 105-108.—Morphology, Natural History. *Scaphiopus.*

323 **Bragg, A. N.** 1956. Further observations on spadefoot toads. Herpetologica 12(3):201-204.—Development, Ecology, Natural History. *Scaphiopus.*

324 **Bragg, A. N.** 1956. Notes on the behavior of toads in captivity. Wasmann J. Biol. 14(2):301-310.—Behavior. *Bufo, Scaphiopus.*

325 **Bragg, A. N.** 1956. In quest of the spadefoots. N. M. Quart. 25(4): 345-357,393.—Behavior, Natural History, Popular. *Scaphiopus.*

326 **Bragg, A. N.** 1956. Spadefoot toads. Okla. Quart. 5(3):29-32.—Natural History, Popular. *Scaphiopus.*

327 **Bragg, A. N.** 1957. Aggregational feeding and metamorphic aggregations in tadpoles of *Scaphiopus hurteri,* observed in 1954. Wasmann J. Biol. 15(1):61-68.—Behavior, Natural History. *Scaphiopus.*

328 **Bragg, A. N.** 1957. Amphibian eggs produced singly or in masses. Herpetologica 13(3):212.—Natural History. *Bufo, Pseudacris, Scaphiopus, Ambystoma.*

329 **Bragg, A. N.** 1957. Some factors in the feeding of toads. Herpetologica 13(3):189-191.—Behavior, Natural History. *Bufo, Gastrophryne, Scaphiopus, Sceloporus.*

330 **Bragg, A. N.** 1957. Use of carrion by the beetle, *Canthon laevis* (Coleoptera, Scarabaeidae). Southwest. Nat. 2(4):173.—The carrion was decaying eastern spadefoot tadpoles. Natural History. *Scaphiopus.*

331 **Bragg, A. N.** 1957. Variation in colors and color patterns in tadpoles of *Scaphiopus bombifrons.* Copeia 1957(1):36-39.—Morphology. *Bufo, Acris, Pseudacris, Gastrophryne, Scaphiopus, Rana.*

332 **Bragg, A. N.** 1958. The clasping reflex observed in juvenile spadefoots. Southwest. Nat. 3(1-4):229.—Behavior. *Scaphiopus.*

333 **Bragg, A. N.** 1958. The eggs of the dwarf American toad. Herpetologica 13(4):273-275.—Natural History. *Bufo.*

334 **Bragg, A. N.** 1958. A melanistic tendency in the Great Plains toad, *Bufo cognatus.* Southwest. Nat. 3(1-4):229-230.—Morphology. *Bufo.*

335 **Bragg, A. N.** 1958. On metamorphic and postmetamorphic aggregations in spadefoots. Herpetologica 13(4):273.—Behavior, Development, Natural History. *Scaphiopus.*

336 **Bragg, A. N.** 1958. Taxonomic status of the gray tree frog in Oklahoma. Herpetologica 14(2):79-84.—Distribution, Taxonomy. *Hyla.*

337 **Bragg, A. N.** 1959. Behavior of tadpoles of Hurter's spadefoot during and exceptionally rainy season. Wasmann J. Biol. 17(1):23-42.—Behavior. *Scaphiopus.*

338 **Bragg, A. N.** 1959. Response of a female *Pseudacris nigrita triseriata* to the call of a male. Copeia 1959(4):341.—Behavior. *Pseudacris.*

339 **Bragg, A. N.** 1959. Voice in Salientia. I. Observations on *Hyla versicolor x chrysoscelis* intergrades in central Oklahoma with a few data from Arkansas. Wasmann J. Biol. 17(2):231-243.—Behavior. *Hyla.*

340 **Bragg, A. N.** 1960. Experimental observations on the feeding of spadefoot tadpoles. Southwest. Nat. 5(4):201-207.—Behavior. *Scaphiopus.*

341 **Bragg, A. N.** 1960. Is *Heterodon* venomous? Herpetologica 16(2): 121-123.—Behavior, Natural History, Health. *Heterodon.*

342 Bragg, A. N. 1960. Observations on Oklahoma toads. Southwest. Nat. 5(2):106.—Journal incorrectly indicates Bragg as author, see King (1960) for correct citation.

343 Bragg, A. N. 1960. Ovulation and breeding patterns in Salientia. Herpetologica 16(2):124.—Ecology, Physiology. *Bufo, Pseudacris, Gastrophryne, Scaphiopus.*

344 Bragg, A. N. 1960. Population fluctuation in the amphibian fauna of Cleveland County, Oklahoma during the past twenty-five years. Southwest. Nat. 5(3):165-169.—Ecology. *Bufo, Acris, Pseudacris, Gastrophryne, Scaphiopus, Rana, Ambystoma.*

345 Bragg, A. N. 1961. The behavior and comparative developmental rates in nature of tadpoles of a spadefoot, a toad, and a frog. Herpetologica 17(2):80-84.—Behavior, Development. *Bufo, Scaphiopus, Rana.*

346 Bragg, A. N. 1961. A theory of the origin of spadefooted toads deduced principally by a study of their habits. Anim. Behav. 9(3-4): 178-186.—Behavior, Evolution. *Scaphiopus.*

347 Bragg, A. N. 1962. Predation on arthropods by spadefoot tadpoles. Herpetologica 18(2):144.—Behavior, Natural History. *Scaphiopus.*

348 Bragg, A. N. 1962. Predator-prey relationship in two species of spadefoot tadpoles with notes on some other features of their behavior. Wasmann J. Biol. 20(1):81-97.—Behavior, Ecology. *Scaphiopus.*

349 Bragg, A. N. 1962. *Saprolegnia* on tadpoles again in Oklahoma. Southwest. Nat. 7(1):79-80.—Parasitology. *Bufo, Pseudacris, Scaphiopus, Rana.*

350 Bragg, A. N. 1963. A hypothesis to explain the almost exclusive use of temporary water by breeding spadefoot toads. Proc. Okla. Acad. Sci. 44:24-25.—Natural History. *Bufo, Hyla, Scaphiopus, Rana.*

351 Bragg, A. N. 1963. Lens action by jelly of frog's eggs. Proc. Okla. Acad. Sci. 44:23.—Natural History. *Rana.*

352 Bragg, A. N. 1964. Further study of predation and cannibalism in spadefoot tadpoles. Herpetologica 20(1):17-24.—Behavior, Natural History. *Scaphiopus.*

353 Bragg, A. N. 1964. Mass movements resulting in aggregations of tadpoles of the Plains spadefoot, some of them in response to light and temperature. (Amphibia; Salientia). Wasmann J. Biol. 22(2):299-305.—Behavior. *Scaphiopus.*

354 **Bragg, A. N.** 1965. Gnomes of the night. Univ. Penn. Press, Philadelphia. 127 pp.—Behavior, Bibliographies, Development, Distribution, Ecology, Evolution, Historical, Lore, Morphology, Natural History, Physiology, Popular, Taxonomy. *Scaphiopus.*

355 **Bragg, A. N.** 1966. Longevity of the tadpole stage in the Plains spadefoot (Amphibia: Salientia). Wasmann J. Biol. 24(1):71-73.—Development. *Scaphiopus.*

356 **Bragg, A. N.** 1966. Rememberances of the Ortenburgers. Biologist 48(1-2):10-14.—General Reference, Historical.

357 **Bragg, A. N.** 1967. Further notes on *Scaphiopus holbrooki hurteri* in Pool P1. Proc. Okla. Acad. Sci. 48:14-15.—Natural History. *Scaphiopus.*

358 **Bragg, A. N.** 1967. Recent studies on the spadefoot toads. Bios 38(2):75-84.—Behavior, Distribution, Ecology, Natural History, Taxonomy. *Scaphiopus.*

359 **Bragg, A. N.** 1968. The formation of feeding schools in tadpoles of spadefoots. Wasmann J. Biol. 26(1):11-16.—Behavior. *Scaphiopus.*

360 **Bragg, A. N.** 1976. An evening with the toads. Bull. Okla. Herp. Soc. 1(3):42-43.—A *post humus* publication prepared by J. H. Black. Behavior, Natural History. *Bufo, Scaphiopus.*

361 **Bragg, A. N., and J. H. Black.** 1968. Four cases of probable albinism in tadpoles. Proc. Okla. Acad. Sci. 49:14-15.—Morphology. *Scaphiopus, Rana.*

362 **Bragg, A. N., and W. N. Bragg.** 1957. The southern painted turtle in Oklahoma. Copeia 1957(4):307-308.—Distribution. *Chrysemys.*

363 **Bragg, A. N., and W. N. Bragg.** 1958. Parasitism of spadefoot tadpoles by *Saprolegnia.* Herpetologica 14(1):34.—Parasitology. *Scaphiopus.*

364 **Bragg, A. N., and W. N. Bragg.** 1958. Variations in the mouth parts in tadpoles of *Scaphiopus* (*Spea*) *bombifrons* Cope (Amphibia: Salientia). Southwest. Nat. 3(1-4):55-69.—Morphology. *Scaphiopus.*

365 **Bragg, A. N., and J. Bresler.** 1951. Viability of the eggs of *Bufo cognatus.* Proc. Okla. Acad. Sci. 32:13-14.—Natural History. *Bufo.*

366 **Bragg, A. N., and M. Brooks.** 1958. Social behavior in juveniles of *Bufo cognatus* Say. Herpetologica 14(3):141-147.—Behavior. *Bufo.*

367 **Bragg, A. N., and V. E. Dowell.** 1954. Leopard frog eggs in October. Proc. Okla. Acad. Sci. 35:41.—Natural History. *Rana.*

368 **Bragg, A. N., and H. A. Dundee.** 1948. Salientian collections in Oklahoma, 1948. Proc. Okla. Acad. Sci. 29:24-25.—Distribution, Natural History. All Anura.

369 **Bragg, A. N., and H. A. Dundee.** 1949. Salientian additions to county lists in Oklahoma. Proc. Okla. Acad. Sci. 30:19-21.—Distribution. *Bufo, Hyla, Pseudacris, Gastrophryne, Scaphiopus, Rana.*

370 **Bragg, A. N., and S. Hayes.** 1963. A study of labial teeth rows in tadpoles of Couch's spadefoot. Wasmann J. Biol. 21(2):149-154.—Morphology. *Scaphiopus.*

371 **Bragg, A. N., and W. F. Hudson.** 1951. A melanistic individual of *Eurycea multiplicata* Cope. Proc. Okla. Acad. Sci. 32:14.—Morphology. *Eurycea.*

372 **Bragg, A. N., and W. F. Hudson.** 1951. New county records of Salientia and a summary of known distribution of Caudata in Oklahoma. Great Basin Nat. 11(3-4):87-90.—Distribution. *Bufo, Pseudacris, Gastrophryne, Scaphiopus, Rana, Ambystoma, Amphiuma, Desmognathus, Eurycea, Plethodon, Typhlotriton, Necturus, Siren, Notophthalmus.*

373 **Bragg, A. N., and O. M. King.** 1960. Aggregational and associated behavior in tadpoles of the Plains spadefoot. Wasmann J. Biol. 18(2): 273-289.—Behavior. *Scaphiopus.*

374 **Bragg, A. N., and R. E. Kuntz.** 1939. *Bufo compactilis* Wiegmann, a new toad for Oklahoma. Proc. Okla. Acad. Sci. 20:77.—Distribution. *Bufo, Pseudacris, Gastrophryne, Scaphiopus, Rana.*

375 **Bragg, A. N., R. Mathews, and R. Kingsinger, Jr.** 1963. The mouth parts of tadpoles of Hurter's spadefoot. Herpetologica 19(4):284-285.—Morphology. *Scaphiopus.*

376 **Bragg, A. N., and J. Nelson.** 1965. Further notes on predation by tadpoles of the Plains spadefoot. Proc. Okla. Acad. Sci. 46:25-26.—Development, Natural History. *Bufo, Scaphiopus.*

377 **Bragg, A. N., and O. Sanders.** 1951. A new subspecies of the *Bufo woodhousei* group of toads (Salientia, Bufonidae). Wasmann J. Biol. 9(3):363-378.—Distribution, Natural History, Taxonomy. *Bufo.*

378 **Bragg, A. N., and C. C. Smith.** 1941. The ecological distribution of toads of the genus *Bufo* in Oklahoma. Bull. Ecol. Soc. Am. 22(4): 30.—An abstract. Distribution, Ecology. *Bufo.*

379 **Bragg, A. N., and C. C. Smith.** 1941. The ecological distribution of toads of the genus *Bufo* in Oklahoma. Anat. Rec. 81(4):70-71.—An abstract. Distribution, Ecology. *Bufo.*

380 **Bragg, A. N., and C. C. Smith.** 1942. Observations on the ecology and natural history of Anura. IX. Notes on breeding behavior in Oklahoma. Great Basin Nat. 3(2):33-50.—Ecology, Natural History. All Anura.

381 **Bragg, A. N., and C. C. Smith.** 1943. New county records of frogs in Oklahoma. Turtox News 21(8):107.—Distribution. *Rana.*

382 **Bragg, A. N., and C. C. Smith.** 1943. Observations on the ecology and natural history of Anura. IV. The ecological distribution of toads in Oklahoma. Ecology 24(3):285-309—Behavior, Distribution, Ecology, Natural History. *Bufo.*

383 **Bragg, A. N., N. Taylor, and R. J. Taylor.** 1968. A range extension of the pickerel frog, *Rana palustris,* in Oklahoma and indications of an extension of the Austroriparian Life Province. Southwest. Nat. 13(3): 372-373.—Distribution. *Rana.*

384 **Bragg, A. N., and A. O. Weese.** 1942. Growth rates and age at sexual maturity of young toads (*Bufo cognatus*) under natural conditions in central Oklahoma. Bull. Ecol. Soc. Am. 23(4):69.—An abstract. Development. *Bufo.*

385 **Bragg, A. N., and A. O. Weese.** 1950. Observations on the ecology and natural history of Anura. XIV. Growth rates and age at sexual maturity of *Bufo cognatus* under natural conditions in central Oklahoma. Pages 47-58 *in* Researches on the Amphibia of Oklahoma. Univ. Okla. Press, Norman.—Behavior, Ecology, Natural History. *Bufo.*

386 **Bragg, A. N., A. O. Weese, H. A. Dundee, H. T. Fisher, A. Richards, and C. B. Clark.** 1950. Researches on the Amphibia of Oklahoma. Univ. Okla. Press, Norman. 154 pp.—Behavior, Distribution, Ecology, Evolution, Morphology, Natural History, Physiology, All Amphibia.

Bragg, A. N., *see* 397, 504, 598, 599, 1317, 1318, 1319, 1320

Bragg, W. N., *see* 362, 363, 364

387 **Brame, A. H., Jr.** 1967. A list of the world's recent and fossil salamanders. Herpeton 2(1):1-26.—Taxonomy. *Eurycea, Plethodon.*

388 **Brandon, R. A.** 1965. *Typhlotriton, T. nereus, T. spelaeus.* Cat. Am. Amphib. Reptiles 20.1-20.2.—Range map and literature review. Bibliographies, Distribution. *Typhlotriton.*

389 **Brandon, R. A.** 1966. A reevaluation of the status of the salamander, *Typhlotriton nereus* Bishop. Copeia 1966(3):555-561.—Distribution, Taxonomy. *Typhlotriton.*

390 **Brandon, R. A.** 1970. *Typhlotriton, T. spelaeus.* Cat. Am. Amphib. Reptiles 84.1-84.2.—Range map and literature review. Bibliographies, Distribution. *Typhlotriton.*

391 **Brandon, R. A., and J. H. Black.** 1970. The taxonomic status of *Typhlotriton braggi* (Caudata, Plethodontidae). Copeia 1970(2):388-391.—Taxonomy. *Typhlotriton.*

Brandon, R. A., *see* 607

392 **Branson, C. C.** 1965. Geology of Byars Fossil Site. Okla. Geol. Notes 25(4):98.—Fossil amphibians are included. General Reference, Paleontology.

Branson, C. C., *see* 1136

393 **Brattstrom, B. H.** 1961. Some fossil tortoises from western North America with remarks on the zoogeography and paleoecology of tortoises. J. Paleontol. 35(3):543-560.—General Reference, Paleontology.

394 **Brattstrom, B. H.** 1967. A succession of Pliocene and Pleistocene snake faunas from the High Plains of the United States. Copeia 1967(1):188-202.—Distribution, Paleontology. *Coluber, Elaphe, Heterodon, Lampropeltis, Nerodia, Pituophis, Thamnophis, Crotalus.*

395 **Bresler, J.** 1951. The development of labial teeth on salientian tadpoles in relation to constant temperature. M.S. Thesis, Univ. Okla., Norman. 48 pp.—Development, Morphology. *Bufo, Rana.*

396 **Bresler, J.** 1954. The development of labial teeth of salientian larvae in relation to temperature. Copeia 1954(3):207-211.—Development. *Bufo, Rana.*

397 **Bresler, J., and A. N. Bragg.** 1954. Variation in the rows of labial teeth in tadpoles. Copeia 1954(4):255-257.—Management. *Bufo, Pseudacris, Scaphiopus, Rana.*

Bresler, J., *see* 365

Britton, J. M., *see* 1459

398 **Brock, B. L.** 1950. Histogenic response to dissimilar growth rates in *Scaphiopus hurteri* Strecker and *Rana pipiens* Schreber. M.S. Thesis, Univ. Okla., Norman. 47 pp.—Development. *Scaphiopus, Rana.*

Brodie, E. D., Jr., *see* 1330

399 **Brooks, F. G.** 1930. Those Oklahoma toads. Biologist 1(3):21-22.—Historical, Popular. *Bufo, Scaphiopus.*

400 **Brooks, G. R.** 1975. *Scincella lateralis.* Cat. Am. Amphib. Reptiles 169.1-169.4.—Range map and literature review. Bibliographies, Distribution. *Scincella.*

401 **Brooks, R. L.** 1987. The Arthur Site: Settlement and subsistence structure at a Washita River phase village. Okla. Archeol. Surv. Bull. (15):1-170.—Box turtle shells found at site. Archeology. *Terrapene.*

Brooks, M., *see* 366

402 **Brooks, R. L., R. R. Drass, and F. E. Swenson.** 1985. Prehistoric farmers of the Washita River Valley: Settlemen and subsistence patterns during the Plains Village Period. Okla. Archeol. Res. Surv. Rep. (23):1-162.—Archeology. *Terrapene.*

Brown, B. C., *see* 1501

Brown, H. P., *see* 1532

Brown, J. A., *see* 1192

403 **Brown, L. E., and J. R. Pierce.** 1967. Male-male interactions and chorusing intensities of the Great Plains toad, *Bufo cognatus.* Copeia 1967(1):149-154.—Behavior, Ecology. *Bufo.*

404 **Brown, M.** 1976. Conservation notebook: Oklahoma's poisonous snakes. Outdoor Okla. 32(6):centerfold.—Popular. *Agkistrodon, Crotalus, Sistrurus.*

405 **Bryce, F. D.** 1975. Reptiles and amphibians of the Wichitas. Great Plains J. 15(1):65-78.—Natural History. *Bufo, Acris, Pseudacris, Gastrophryne, Scaphiopus, Rana, Chelydra, Trachemys, Terrapene, Kinosternon, Trionyx, Ophisaurus, Crotaphytus, Holbrookia, Phrynosoma, Sceloporus, Eumeces, Scincella, Cnemidophorus, Arizona, Coluber, Diadophis, Elaphe, Heterodon, Hypsiglena, Lampropeltis, Masticophis, Nerodia, Opheodrys, Pituophis, Regina, Sonora, Storeria, Tantilla, Thamnophis, Tropidoclonion, Leptotyphlops, Agkistrodon, Crotalus, Sistrurus.*

406 **Bryce, F. D.** 1977. Range records of the western scarlet snake and small-mouthed salamander. Bull. Okla. Herp. Soc. 2(1):14.—Distribution. *Ambystoma, Holbrookia, Sceloporus, Scincella, Cnemidophorus, Cemophora.*

407 **Bryce, F. D.** 1977. Ven-Am educational activities. Bull. Okla. Herp. Soc. 2(2-3):38-39.—Popular. *Bufo, Gastrophryne.*

408 **Bryce, F. D.** 1980. *Coluber c. priapus* in Oklahoma. Bull. Okla. Herp. Soc. 5(1):9.—Distribution. *Coluber.*

Bull, J. J., *see* 84

409 **Burger, W. L., Jr.** 1947. A taxonomic and statistical study of the keeled green snake, *Opheodrys aestivus.* Bull. Ecol. Soc. Am. 28(4): 54.—An abstract. Distribution, Taxonomy. *Opheodrys.*

410 **Burger, W. L., Jr.** 1948. *Farancia abacura reinwardtii* in Oklahoma. Copeia 1948(2):133.—Distribution. *Farancia.*

411 **Burger, W. L., Jr., P. W. Smith, and H. M. Smith.** 1949. Notable records of reptiles and amphibians in Oklahoma, Arkansas and Texas. J. Tenn. Acad. Sci. 24(2):130-134.—Distribution. *Hyla, Gastrophryne, Pseudemys, Kinosternon.*

Burger, W. L., Jr., *see* 600, 865

412 **Burkett, R. D.** 1966. Natural history of cottonmouth moccasin, *Agkistrodon piscivorus* (Reptilia). Univ. Kans. Publ. Mus. Nat. Hist. 17(9):435-491.—Distribution, Morphology. *Crotalus.*

413 **Burt, C. E.** 1928. The synonomy, variation, and distribution of the collared lizard, *Crotaphytus collaris* (Say). Occas. Pap. Mus. Zool. Univ. Mich. (196):1-19.—Distribution, Taxonomy. *Crotaphytus.*

414 **Burt, C. E.** 1929. The synonomy, variation, and distribution of the Sonoran skink, *Eumeces obsoletus* (Baird and Girard). Occas. Pap. Mus. Zool. Univ. Mich. (201):1-12.—Distribution, Morphology, Taxonomy. *Eumeces.*

415 **Burt, C. E.** 1931. A report on some amphibians and reptiles from Kansas, Nebraska, and Oklahoma. Proc. Biol. Soc. Wash. (44):11-16. —Distribution, Taxonomy. *Bufo, Acris, Rana, Chelydra, Terrapene, Crotaphytus, Holbrookia, Scincella, Coluber, Elaphe, Lampropeltis, Pituophis, Thamnophis.*

416 **Burt, C. E.** 1931. A study of the teiid lizards of the genus *Cnemidophorus* with special reference to their phylogenetic relation-

ships. Bull. U. S. Natl. Mus. (154):1-286.—Bibliographies, Distribution, Evolution, Taxonomy. *Cnemidophorus.*

417 **Burt, C. E.** 1932. Records of amphibians from the eastern and central United States (1931). Am. Midl. Nat. 13(2):75-86.—Distribution. *Pseudacris.*

418 **Burt, C. E.** 1935. Further records of the ecology and distribution of amphibians and reptiles in the middle west. Am. Midl. Nat. 16(3): 311-336.—Distribution. All Anura, *Desmognathus, Chelydra, Trachemys, Terrapene, Kinosternon, Ophisaurus, Crotaphytus, Holbrookia, Phrynosoma, Sceloporus, Eumeces, Scincella, Cnemidophorus, Arizona, Coluber, Diadophis, Elaphe, Heterodon, Lampropeltis, Masticophis, Nerodia, Opheodrys, Pituophis, Regina, Sonora, Tantilla, Tropidoclonion, Virginia.*

419 **Burt, C. E.** 1935. A key to the lizards of the United States and Canada. Trans. Kans. Acad. Sci. 38:255-305.—Distribution, Morphology, Taxonomy. All Sauria.

420 **Burt, C. E.** 1938. Contributions to Texan herpetology VII. The salamanders. Am. Midl. Nat. 20(2):374-380.—Mentions mudpuppies in Oklahoma. Distribution. *Necturus.*

421 **Burt, C. E., and M. D. Burt.** 1929. A collection of amphibians and reptiles from the Mississippi Valley, with field observations. Am. Mus. Novit. (381):1-14.—Specimens were collected in Oklahoma during the summer of 1923. Distribution. *Acris, Rana, Terrapene, Crotaphytus, Phrynosoma, Cnemidophorus, Coluber, Lampropeltis, Regina, Tantilla.*

422 **Burt, C. E., and M. D. Burt.** 1929. Field notes and locality records on a collection of amphibians and reptiles chiefly from the western half of the United States. I. Amphibians. II. Reptiles. J. Wash. Acad. Sci. 19:428-434.—Distribution. *Acris, Rana, Terrapene, Crotaphytus, Coluber, Nerodia*

423 **Burt, C. E., and W. L. Hoyle.** 1935. Additional records of the reptiles of the central prairie region of the United States. Trans. Kans. Acad. Sci. 37:193-216.—Distribution. *Chrysemys, Terrapene, Kinosternon, Crotaphytus, Phrynosoma, Sceloporus, Eumeces, Scincella, Cnemidophorus, Coluber, Diadophis, Elaphe, Lampropeltis, Masticophis, Pituophis, Sonora, Tantilla, Thamnophis.*

Burt, M. D., *see* 421, 422

424 **Bury, R. B., C. K. Dodd, Jr., and G. M. Fellers.** 1980. Conservation of the Amphibia of the United States: A review. U. S. Dept. Inter. Fish Wildl. Serv. Res. Publ. (134):1-34.—Management. *Eurycea, Plethodon.*

Buscher, H. N., *see* 1442

Butcher, M., *see* 150, 151, 975

425 **Butler, B. H.** 1977. Section IV: Analysis of the faunal materials from the Big Hawk Shelter. Pages 57-87 *in* D. O. Henry (Ed.). The prehistory and paleoenvironment of Hominy Creek Valley. Univ. Tulsa. 206 pp.—Archeology. *Rana, Trachemys, Terrapene, Trionyx, Nerodia, Thamnophis, Agkistrodon, Crotalus.*

426 **Butler, B. H.** 1978. Vertebrate faunal remains. Pages 49-78 *in* D. O. Henry (Ed.). The prehistory and paleoenvironment of Hominy Creek Valley. Univ. Tulsa. 102 pp.—Archeology. *Rana, Graptemys, Trachemys, Terrapene, Trionyx, Elaphe, Nerodia, Thamnophis, Agkistrodon, Crotalus.*

Butler, B. H., *see* 756

Byers, E., *see* 1071

427 **Byrd, K. M.** 1983. Zooarcheological analysis, Bug Hill: Excavation of the multicomponent midden mound in the Jackfork Valley, southeast Oklahoma. Pages 220-232 *in* J. H. Altschul (Ed.). New World Res. Rept. Invest. 81-1, Pollock, Louisiana.—Archeology. *Rana, Chelydra, Chrysemys, Graptemys, Trachemys, Terrapene, Kinosternon, Trionyx.*

428 **Cagle, F. R.** 1953. Two new subspecies of *Graptemys pseudogeographica.* Occas. Pap. Mus. Zool. Univ. Mich. (546):1-17.—Distribution, Taxonomy. *Graptemys.*

429 **Calloway, A.** 1973. Oklahoma's snapper capitol. Outdoor Okla. 29(7):22-23.—Photograph of a large turtle. Popular. *Macroclemys.*

430 **Capron, L. A.** 1976. Rattlesnake hunts: Crimes against nature. Kans. Herpetol. Soc. Newsl. (13):4.—Criticism of Oklahoma snake hunts. Management. *Crotalus.*

431 **Capron, M.** 1985. A western diamondback rattlesnake released in Sumner County, Kansas. Kans. Herpetol. Soc. Newsl. (59):5-6.—Oklahoma snake released in Kansas. Distribution. *Crotalus.*

432 **Carpenter, C. C.** 1954. Feeding aggregation of narrow mouth toads (*Microhyla carolinensis olivacea*). Proc. Okla. Acad. Sci. 35:45.—Behavior, Natural History. *Gastrophryne.*

433 **Carpenter, C. C.** 1954. Records of distribution for amphibians and reptiles of Oklahoma. Proc. Okla. Acad. Sci. 35:39-41.—Distribution. *Hyla, Gastrophryne, Scaphiopus, Rana, Ambystoma, Necturus, Siren, Notophthalmus, Deirochelys, Terrapene, Sternotherus, Ophisaurus, Anolis, Crotaphytus, Phrynosoma, Eumeces, Cemophora, Lampropeltis, Nerodia, Opheodrys, Regina, Rhinocheilus, Storeria, Thamnophis, Agkistrodon.*

434 **Carpenter, C. C.** 1954. An unusually large diamond-backed water snake (*Natrix rhombifera*) from Buncombe Creek, Lake Texoma. Proc. Okla. Acad. Sci. 35:42.—Natural History. *Nerodia.*

435 **Carpenter, C. C.** 1955. The amphibians and reptiles of the University of Oklahoma Biological Station area in south central Oklahoma. Proc. Okla. Acad. Sci. 36:39-46.—Bibliographies, Distribution. All Anura, *Ambystoma, Siren, Notophthalmus, Ophisaurus, Phrynosoma, Sceloporus, Eumeces, Scincella, Cnemidophorus, Coluber, Diadophis, Elaphe, Heterodon, Lampropeltis, Masticophis, Nerodia, Opheodrys, Pituophis, Sonora, Storeria, Tantilla, Thamnophis, Tropidoclonion, Virginia, Agkistrodon, Crotalus, Sistrurus.*

436 **Carpenter, C. C.** 1955. Sounding turtles: A field locating technique. Herpetologica 11(2):120.—Techniques. *Chelydra, Kinosternon, Trachemys.*

437 **Carpenter, C. C.** 1956. Carapace pits in the three-toed box turtle, *Terrapene carolina triunguis* (Chelonia-Emydidae). Southwest. Nat. 1(2):83-86.—Morphology. *Terrapene.*

438 **Carpenter, C. C.** 1956. Reptiles and amphibians of the Oliver Wildlife Preserve. Proc. Okla. Acad. Sci. 37:33-34.—Natural History. *Bufo, Acris, Pseudacris, Gastrophryne, Rana, Ambystoma, Chelydra, Trachemys, Terrapene, Kinosternon, Ophisaurus, Sceloporus, Eumeces, Scincella, Cnemidophorus, Coluber, Diadophis, Elaphe, Lampropeltis, Masticophis, Opheodrys, Regina, Storeria, Thamnophis.*

439 **Carpenter, C. C.** 1957. Additional distribution records for Oklahoma reptiles. Proc. Okla. Acad. Sci. 38:71-74.—Distribution. *Chelydra, Pseudemys, Trachemys, Terrapene, Kinosternon, Sternotherus, Trionyx, Ophisaurus, Crotaphytus, Holbrookia, Phrynosoma, Sceloporus, Eumeces, Scincella, Cnemidophorus, Carphophis, Coluber, Diadophis, Elaphe, Heterodon, Hypsiglena, Lampropeltis, Masticophis, Nerodia, Opheodrys, Pituophis, Regina, Sonora, Storeria, Tantilla, Thamnophis, Tropidoclonion, Virginia, Agkistrodon, Crotalus.*

440 Carpenter, C. C. 1957. An albino speckled kingsnake. Herpetologica 13(1):78.—Morphology. *Lampropeltis.*

441 Carpenter, C. C. 1957. Hibernation, hibernacula and associated behavior of the three-toed box turtle (*Terrapene carolina triunguis*). Copeia 1957(4):278-282.—Behavior, Ecology. *Chelydra, Trachemys, Terrapene, Kinosternon.*

442 Carpenter, C. C. 1958. Reproduction, young, eggs and food of Oklahoma snakes. Herpetologica 14(2):113-115.—Natural History. *Coluber, Elaphe, Heterodon, Lampropeltis, Masticophis, Nerodia, Opheodrys, Pituophis, Sonora, Storeria, Tantilla, Thamnophis, Virginia, Agkistrodon, Crotalus.*

443 Carpenter, C. C. 1958. An unusual Ouachita map turtle. Herpetologica 14(2):116.—Morphology. *Graptemys.*

444 Carpenter, C. C. 1959. Checklist of amphibians and reptiles. Pages 25-29 *in* Guide to Roman Nose State Park. Okla. Geol. Surv. Guidebook 9. 31 pp.—Distribution. *Bufo, Acris, Pseudacris, Gastrophryne, Scaphiopus, Rana, Ambystoma, Chelydra, Trachemys, Terrapene, Kinosternon, Trionyx, Crotaphytus, Phrynosoma, Sceloporus, Eumeces, Scincella, Cnemidophorus, Arizona, Coluber, Elaphe, Heterodon, Hypsiglena, Lampropeltis, Masticophis, Nerodia, Opheodrys, Pituophis, Sonora, Storeria, Tantilla, Thamnophis, Agkistrodon, Crotalus, Sistrurus.*

445 Carpenter, C. C. 1959. A population of *Sceloporus undulatus consobrinus* in south-central Oklahoma. Southwest. Nat. 4(2):110-111.—Behavior, Ecology. *Sceloporus.*

446 Carpenter, C. C. 1959. A population of the six-lined racerunner (*Cnemidophorus sexlineatus*). Herpetologica 15(2):81-86.—Behavior, Ecology, Techniques. *Cnemidophorus.*

447 Carpenter, C. C. 1959. Reptile cents. Okla. Wildl. 15(3):14.—General Reference, Popular.

448 Carpenter, C. C. 1960. Aggressive behavior and social dominance in the six-lined racerunner (*Cnemidophorus sexlineatus*). Anim. Behav. 8(1-2):61-66.—Behavior, Techniques. *Cnemidophorus.*

449 Carpenter, C. C. 1960. A large brood of western pigmy rattlesnakes. Herpetologica 16(2):142-143.—Natural History. *Sistrurus.*

450 Carpenter, C. C. 1960. Reproduction in Oklahoma *Sceloporus* and *Cnemidophorus.* Herpetologica 16(3):175-182.—Natural History, Techniques. *Sceloporus, Cnemidophorus.*

451 **Carpenter, C. C.** 1960. Temperature relationships of two Oklahoma lizards. Proc. Okla. Acad. Sci. 41:72-77.—Behavior, Ecology. *Sceloporus, Cnemidophorus.*

452 **Carpenter, C. C.** 1963. Amphibians and reptiles of McCurtain County. Pages 33-34 *in* Guide to Beavers Bend State Park. Okla. Geol. Surv. Guidebook 11. 46 pp.—Distribution. All Anura, *Ambystoma, Desmognathus, Eurycea, Plethodon, Siren, Notophthalmus,* All Testudines, *Ophisaurus, Anolis, Crotaphytus, Sceloporus, Eumeces, Scincella, Cnemidophorus, Carphophis, Coluber, Diadophis, Elaphe, Farancia, Lampropeltis, Masticophis, Nerodia, Opheodrys, Regina, Sonora, Storeria, Tantilla, Thamnophis, Virginia, Agkistrodon, Crotalus.*

453 **Carpenter, C. C.** 1963. Revision of checklist of amphibians and reptiles of the Wichita Mountains Wildlife Refuge. Mimeographed handout printed at the Refuge.—Distribution. *Bufo, Acris, Pseudacris, Gastrophryne, Scaphiopus, Rana, Ambystoma, Chelydra, Graptemys, Trachemys, Terrapene, Kinosternon, Trionyx, Ophisaurus, Crotaphytus, Holbrookia, Phrynosoma, Sceloporus, Eumeces, Scincella, Cnemidophorus, Arizona, Coluber, Diadophis, Elaphe, Heterodon, Hypsiglena, Lampropeltis, Masticophis, Nerodia, Opheodrys, Pituophis, Regina, Rhinocheilus, Sonora, Storeria, Tantilla, Thamnophis, Tropidoclonion, Leptotyphlops, Agkistrodon, Crotalus, Sistrurus.*

454 **Carpenter, C. C.** 1966. Patterns of behavior in two Oklahoma lizards. Am. Midl. Nat. 67(1):132-151.—Behavior. *Sceloporus, Cnemidophorus.*

455 **Carpenter, C. C.** 1968. Social structure and display patterns of iguanid lizards. Exercise 24. Pages 109-112 *in* A. W. Stokes (Ed.). Animal behavior in laboratory and field. W. H. Freeman and Co., San Fransisco. 819 pp.—Behavior. *Anolis, Crotaphytus, Holbrookia, Phrynosoma, Sceloporus, Uta.*

456 **Carpenter, C. C.** 1972. Herpetofauna of the Kiamichi Basin. Pages 26-39 *in* Analysis of the biology of the Kiamichi River, Oklahoma. Univ. Okla. Biol. Surv. 147 pp.—Distribution. All Anura, *Ambystoma, Desmognathus, Eurycea, Plethodon, Siren, Notophthalmus, Chelydra, Macroclemys, Chrysemys, Deirochelys, Graptemys, Pseudemys, Terrapene, Kinosternon, Sternotherus, Trionyx, Ophisaurus, Anolis, Crotaphytus, Phrynosoma, Sceloporus, Eumeces, Scincella, Cnemidophorus, Carphophis, Cemophora, Coluber, Diadophis, Elaphe, Heterodon, Lampropeltis, Masticophis, Nerodia, Opheodrys, Regina, Sonora, Storeria, Tantilla, Thamnophis, Tropidoclonion, Virginia, Agkistrodon, Crotalus, Sistrurus, Alligator.*

457 **Carpenter, C. C.** 1975. Reptiles and amphibians of the Sooner Site Study Area. Vol. I. Terrestrial field studies. 3.5:1-3.5:31. Environ. Assess. Rep., Sooner Gen. Station, Okla. Gas Elec. Co. Semiannual Rep. (Feb-July), Beham-Blair and Affiliates, Inc.—Distribution, Ecology. Techniques, All Anura, *Chelydra, Trachemys, Terrapene, Kinosternon, Ophisaurus, Crotaphytus, Phrynosoma, Eumeces, Scincella, Cnemidophorus, Coluber, Diadophis, Elaphe, Lampropeltis, Masticophis, Nerodia, Opheodrys, Pituophis, Storeria, Tantilla, Thamnophis, Tropidoclonion, Virginia, Agkistrodon, Crotalus.*

458 **Carpenter, C. C.** 1977. Communication and displays of snakes. Am. Zool. 17(1):217-223.—Behavior. *Heterodon, Agkistrodon, Crotalus, Sistrurus.*

459 **Carpenter, C. C.** 1977. DOKARRS-Distribution of Oklahoma amphibians and reptiles by recorded sightings. Bull. Okla. Herp. Soc. 2(1):22-25.—Distribution, Techniques.

460 **Carpenter, C. C.** 1977. History of Oklahoma Herpetology. II. Dr. Arthur Irving Ortenburger 1898-1961. Bull. Okla. Herp. Soc. 2(2-3): 52-55.—Bibliographies, General Reference, Historical.

461 **Carpenter, C. C.** 1978. Comparative display behavior in the genus *Sceloporus* (Iguanidae). Milw. Publ. Mus. Contrib. Biol. Geol. (18):1-71.—Behavior, Taxonomy, Techniques. *Sceloporus.*

462 **Carpenter, C. C.** 1978. Ecology, terrestrial life resources, reptiles and amphibians. Pages 2:80-2:84 *in* Environ. Assess. Rep., Okla. Grand River Dam Authority, 490-MW Coal-Fired Gen. Station, Unit No. 1 Vol. 1.—Distribution. *Gastrophryne, Chelydra, Graptemys, Trachemys, Terrapene, Kinosternon, Ophisaurus, Eumeces, Scincella, Nerodia, Tropidoclonion.*

463 **Carpenter, C. C.** 1978. Ritualistic social behavior in lizards. Pages 253-267 *in* N. Greenberg and P. Maclean (Eds.). Behavior and neurology of lizards: An interdisciplinary colloquim. U.S. Dept. Health Ed. Welfare, Nat. Inst. Mental Health, Rockville, Md. 352 pp.—Behavior, Bibliographies. *Crotaphytus, Holbrookia, Sceloporus, Cnemidophorus.*

464 **Carpenter, C. C.** 1978. 7-footed barred tiger salamander. Bull. Okla. Herp. Soc. 3(4):65-67.—Morphology. *Ambystoma.*

465 **Carpenter, C. C.** 1979. A combat ritual between two male pygmy rattlesnakes (*Sistrurus miliarius*). Copeia 1979(4):638-642.—Behavior. *Sistrurus.*

466 **Carpenter, C. C.** 1979. Photographic analysis. Pages 97-105 *in* M. Harless and H. Morlock (Eds.). Turtles: Prespective and research. John Wiley and Sons, New Jersey.—Techniques. *Terrapene.*

467 **Carpenter, C. C.** 1980. An ethological approach to reproductive success in reptiles. Pages 33-48 *in* J. B. Murphy and J. T. Collins (Eds.). Reproductive biology and diseases of captive reptiles. Soc. Study Amphib. Reptiles Contrib. Herpetol. (1):1-277.—Behavior, Natural History. *Kinosternon, Trionyx, Scincella, Cnemidophorus.*

468 **Carpenter, C. C.** 1980. Further distribution records for Oklahoma reptiles. Bull. Okla. Herp. Soc. 4(4):66-69.—Distribution. *Chelydra, Graptemys, Trachemys, Terrapene, Kinosternon, Trionyx, Ophisaurus, Anolis, Phrynosoma, Sceloporus, Eumeces, Scincella, Arizona, Coluber, Diadophis, Elaphe, Heterodon, Hypsiglena, Lampropeltis, Masticophis, Opheodrys, Sonora, Storeria, Tantilla, Thamnophis, Tropidoclonion, Virginia, Leptotyphlops, Agkistrodon, Crotalus.*

469 **Carpenter, C. C.** 1980. Southern Oklahoma herpetological distribution records. Bull. Okla. Herp. Soc. 5(3):47-48.—Distribution. *Ambystoma, Notophthalmus, Graptemys, Trachemys, Eumeces, Coluber, Storeria.*

470 **Carpenter, C. C.** 1981. *Trionyx spiniferus* (spiny softshell). Morphology. Herpetol. Rev. 12(3):82.—Development, Natural History. *Trionyx.*

471 **Carpenter, C. C.** 1981. Two intensive herpetological surveys in Oklahoma. Bull. Okla. Herp. Soc. 6(2):26-38.—Distribution, Ecology, Techniques. All Anura, *Chelydra, Chrysemys, Trachemys, Terrapene, Kinosternon, Ophisaurus, Crotaphytus, Phrynosoma, Sceloporus, Scincella, Cnemidophorus, Carphophis, Coluber, Diadophis, Elaphe, Lampropeltis, Masticophis, Nerodia, Opheodrys, Pituophis, Regina, Sonora, Storeria, Tantilla, Thamnophis, Tropidoclonion, Virginia, Agkistrodon, Crotalus.*

472 **Carpenter, C. C.** 1982. The bullsnake as an excavator. J. Herpetol. 16(4):394-401.—Behavior. *Pituophis.*

473 **Carpenter, C. C.** 1982. DOKARRS request. Bull. Okla. Herp. Soc. 7(1):13-16.—General Reference, Popular, Techniques.

474 **Carpenter, C. C.** 1982. A new western record for the alligator snapping turtle (*Macroclemys temmincki*) in Oklahoma. Bull. Okla. Herp. Soc. 7(1):2-3.—Distribution. *Macroclemys.*

475 **Carpenter, C. C.** 1984. Enclosure maintenance of lizards. Pages 69-76 *in* V. L. Bels and A. P. Van den Sande (Eds.). Acta zoologica et pathologica antverpiensia; Maintenance and reproduction of reptiles in captivity, Vol. I.—General Reference, Techniques.

476 **Carpenter, C. C.** 1985. *Lampropeltis calligaster calligaster* (prairie kingsnake). Reproduction. Herpetol. Rev. 16(3):81.—Natural History. *Lampropeltis.*

477 **Carpenter, C. C.** 1985. *Tantilla gracilis* (flat-headed snake). Albino. Herpetol. Rev. 16(3):81.—Morphology. *Tantilla.*

478 **Carpenter, C. C., H. L. Cleveland, W. A. Talbot, J. M. Goode, H. J. Merriman, and D. R. Taylor.** 1961. A *Uta* invasion of Oklahoma. Southwest. Nat. 6(3-4):192-193.—Distribution. *Uta.*

479 **Carpenter, C. C., and J. C. Gillingham.** 1975. Postural responses to kingsnakes by crotaline snakes. Herpetologica 31(3):293-302.—Behavior. *Lampropeltis, Agkistrodon, Crotalus, Sistrurus.*

480 **Carpenter, C. C., and G. Grubitz, III.** 1961. Time-motion study of a lizard. Ecology 42(1):199-200.—Behavior, General Reference, Techniques.

481 **Carpenter, C. C., and J. K. Yoshida.** 1966. Predation on lizards by *Callisoma* (Coleoptera). Southwest. Nat. 11(3):413-414.—Natural History. *Holbrookia.*

Carpenter, C. C., *see* 152, 523, 680, 685, 853, 958, 1214, 1215, 1216, 1276, 1287

482 **Carpenter, G. C., and D. Duvall.** 1983. Motion picture and videotape analysis of behavior. Am. Biol. Teach. 45(6):349-352.—General Reference, Techniques.

483 **Carpenter, J. R.** 1939. Ecology of the grassland biome. Ph.D. Diss., Univ. Okla., Norman. 215 pp.—Distribution. *Bufo, Terrapene, Phrynosoma, Eumeces, Pituophis.*

484 **Carr, A. F., Jr.** 1938. A new subspecies of *Pseudemys floridana,* with notes on the *floridana* complex. Copeia 1938(3):105-109.—Distribution, Morphology, Taxonomy. *Pseudemys.*

485 **Carr, A. F., Jr.** 1949. The identity of *Malacoclemmys kohnii* Baur. Herpetologica 5(1):9-10.—Distribution, Taxonomy. *Graptemys.*

486 **Carr, A. F., Jr.** 1952. Handbook of turtles: The turtles of the United States, Canada, and Baja California. Comstock Publ. Co., Ithaca. 542 pp.—Distribution. All Testudines.

Carr, C. A., *see* 1169

487 **Carroll, R. L.** 1968. A ?diapsid (Reptilia) parietal from the Lower Permian of Oklahoma. Postilla (117):1-7.—General Reference, Paleontology.

Carroll, R. L., *see* 505

488 **Carter, W. A.** 1965. Distribution records for Oklahoma reptiles. Proc. Okla. Acad. Sci. 46:33-36.—Distribution. *Chelydra, Graptemys, Trachemys, Terrapene, Kinosternon, Sternotherus, Trionyx, Ophisaurus, Crotaphytus, Phrynosoma, Sceloporus, Eumeces, Scincella, Cnemidophorus, Carphophis, Cemophora, Diadophis, Elaphe, Heterodon, Hypsiglena, Lampropeltis, Masticophis, Nerodia, Opheodrys, Pituophis, Regina, Storeria, Tantilla, Tropidoclonion, Virginia, Leptotyphlops, Agkistrodon, Crotalus, Sistrurus.*

489 **Carter, W. A.** 1968. First record of *Hemidactylium scutatum* in Oklahoma. Herpetologica 24(4):328.—Distribution, Ecology. *Hemidactylium.*

490 **Carter, W. A., and R. Cox.** 1966. Amphibians and reptiles known from Pontotoc County, Oklahoma. Proc. Okla. Acad. Sci. 47:66-71.—Distribution. All Anura, *Ambystoma, Chelydra, Macroclemys, Pseudemys, Trachemys, Terrapene, Kinosternon, Sternotherus, Trionyx, Ophisaurus, Crotaphytus, Phrynosoma, Sceloporus, Eumeces, Scincella, Cnemidophorus, Coluber, Diadophis, Elaphe, Heterodon, Hypsiglena, Lampropeltis, Masticophis, Nerodia, Opheodrys, Pituophis, Sonora, Storeria, Tantilla, Thamnophis, Tropidoclonion, Virginia, Leptotyphlops, Agkistrodon, Crotalus, Sistrurus.*

Carter, W. J., *see* 1072

491 **Cartledge, T. R.** 1970. The Tyler-Rose Site and late prehistory in east central Oklahoma. Okla. River Basin Surv. Archeol. Site Rep. 19:1-82.—Both fossil amphibians and reptiles are mentioned. Archeology, General Reference.

492 **Case, E. C.** 1902. On some vertebrates from the Permian beds of Oklahoma. Okla. Dept. Geol. Nat. Hist. 2nd. Biann. Rep. Pages 62-68. Both fossil amphibians and reptiles are mentioned. General Reference, Paleontology.

493 **Case, E. C.** 1907. Revision of the Pelycosauria of North America. Carnegie Inst. Wash. Publ. (55):1-176.—Fossil reptiles collected in the Indian Territory that is now Oklahoma. General Reference, Paleontology.

494 **Case, E. C.** 1915. The Permo-Carboniferons red beds of North America and their vertebrate fauna. Carnegie Inst. Wash. Publ. (207):1-176.—Both fossil amphibians and reptiles from Oklahoma are included. Distribution, General Reference, Paleontology.

495 **Case, E. C., E. H. Sellards, and L. Hussakof.** 1911. Revision of the Amphibia and Pisces of the Permian of North America. Carnegie Inst. Wash. Publ. (146):1-179.—General Reference, Paleontology.

Castillo, R. E., *see* 1360

496 **Cathey, T. M.** 1983. Southern leopard frogs in the diet of a racoon. Bull. Okla. Herp. Soc. 8(2):54.—Natural History. *Rana.*

Cathey, T. A., *see* 797

497 **Chantell, C. J.** 1966. Late Coenozoic hylids from the Great Plains. Herpetologica 22(4):259-264.—Paleontology. *Acris, Hyla, Pseudacris.*

498 **Chase, H. D., and A. P. Blair.** 1937. Two new blind isopods from northeastern Oklahoma. Am. Midl. Nat. 18(2):220-224.—Salamander larvae were in association with the isopods in a cave near Kansas, Oklahoma. Ecology. *Eurycea, Typhlotriton.*

Chase, H. D., *see* 180

499 **Chenoweth, W. L.** 1948. Birth and behavior of young copperheads. Herpetologica 4(5):162.—Behavior, Natural History. *Agkistrodon.*

500 **Chichester, H. G.** 1978. Gas transport of aquatic and terrestrial turtle bloods. M.S. Thesis, Univ. Okla., Norman. 35 pp.—Physiology. *Chelydra, Trachemys, Terrapene, Trionyx.*

Chorn, J., *see* 1274

501 **Chrapliwy, P. S.** 1953. Taxonomy and distribution of the spadefoot toads of North America (Salientia: Pelobatidae). M.S. Thesis, Univ. Kansas, Lawrence. 131 pp.—Distribution, Taxonomy. *Scaphiopus.*

Christensen, M. T., *see* 579, 895

502 **Christman, S. P.** 1982. *Storeria dekayi.* Cat. Am. Amphib. Reptiles 306.1-306.4.—Range map and literature review. Bibliographies, Distribution. *Storeria.*

503 **Claborn, A. H.** 1983. A comparative review of the life history of *Graptemys ouchatensis* and *Chrysemys scripta elegans.* M.N.S. Thesis, Univ. Okla., Norman. 35 pp.—Ecology, Natural History. *Graptemys, Trachemys.*

504 **Clark, C. B., and A. N. Bragg.** 1950. A comparison of the ovaries of two species of *Bufo* with different ecological requirements. Pages 143-152 *in* Researches on the Amphibia of Oklahoma. Univ. Okla. Press, Norman.—Behavior, Ecology, Morphology, Physiology. *Bufo.*

Clark, C. B., *see* 386

505 **Clark, J., and R. L. Carroll.** 1973. Romeriid reptiles from the Lower Permian. Bull. Mus. Comp. Zool. 144(5):353-406.—General Reference, Paleontology.

506 **Clark, V. S.** 1977. A report on rattlesnake roundups in Oklahoma, and other parts of the United States. Bull. Okla. Herp. Soc. 2(4):70-80.—Management. *Crotalus.*

507 **Clarke, R. F.** 1963. The comparative behavior of the iguanid lizards, *Callisaurus, Cophosaurus,* and *Holbrookia.* Ph.D. Diss., Univ. Okla., Norman. 136 pp.—Behavior, Ecology, Morphology, Taxonomy. *Holbrookia.*

508 **Clarke, R. F.** 1965. An ethological study of the iguanid genera *Callisaurus, Cophosaurus,* and *Holbrookia.* Emporia State Res. Stud. 13(4):1-66.—Behavior, Ecology, Morphology, Taxonomy. *Holbrookia.*

509 **Clarke, R. F.** 1970. Salamanders in Kansas and vicinity. Kans. School Nat. 16(4):1-16.—Distribution, Ecology, Natural History. *Ambystoma, Desmognathus, Eurycea, Plethodon, Typhlotriton, Necturus, Notophthalmus.*

510 **Clarke, R. F.** 1972. Discussion. Pages 96-96 *in* R. T. Allen and F. C. James (Eds.). A symposium on ecosystematics. Occas. Pap. Univ. Arkansas Mus. (4):235.—Discussion on life history differences in lizards. Ecology. *Sceloporus.*

511 **Clarke, R. F.** 1987. Second report of *Phrynosoma modestum* in Oklahoma. Bull. Okla. Herp. Soc. 12(1-4):16—Distribution, Ecology. *Crotaphytus, Holbrookia, Sceloporus, Cnemidophorus.*

512 **Clary, J. W.** 1941. The trematodes of reptiles, Part I. Systematic section. M.S. Thesis, Okla. State Univ., Stillwater.—Only available as a bound reprint of Hughes, Higgins, and Clary. 1942. Am. Midl. Nat. 27(1):109-134. Parasitology. *Trachemys, Heterodon.*

Clary, J. W., *see* 798

513 **Clay, W. M.** 1938. A synopsis of the North American water snakes of the genus *Natrix.* Copeia 1938(4):173-182.—Mentions distribution of Oklahoma forms. Distribution. *Nerodia.*

514 **Clemons, M. D.** 1978. Ocular blood discharge by *Phrynosoma.* Bull. Okla. Herp. Soc. 3(1):14.—Natural History. *Phrynosoma.*

515 **Clemons, M. K.** 1977. Herptiles and fishes active under ice. Bull. Okla. Herp. Soc. 2(1):18.—Natural History. *Rana, Trachemys.*

Cleveland, H. L., *see* 478

Cline, B. A., *see* 517

516 **Cline, G. R.** 1986. Miscellaneous accounts of some herptiles of Ottawa County, Oklahoma. Bull. Okla. Herp. Soc. 11(1-4):32-33.—Behavior, Distribution, Ecology, Natural History. *Hyla, Ambystoma, Trachemys, Eumeces.*

517 **Cline, G. R., and B. A. Cline.** 1986. Etymology of mountain boomer. Bull. Okla. Herp. Soc. 11(1-4):16-19.—Behavior, Historical, Lore, Popular. *Crotaphytus.*

518 **Cline, G. R., D. R. Edds, and D. A. Rutherford.** 1983. *Scincella lateralis* (Say) in Jefferson County, Oklahoma. Bull. Okla. Herp. Soc. 8(4): 100-101.—Distribution. *Scincella.*

Cline, G. R., *see* 10, 854

519 **Cochran, D. M.** 1961. Type specimens of reptiles and amphibians in the United States National Museum. Bull. U.S. Natl. Mus. (220):1-291.—Distribution, Taxonomy. *Bufo, Eurycea, Plethodon, Holbrookia, Arizona, Thamnophis.*

520 **Collins, J. T.** 1974. Amphibians and reptiles in Kansas. Univ. Kans. Nat. Hist. Publ. Educ. Sec. (1):1-281.—Author makes reference to observations on Oklahoma specimens. Distribution. *Bufo, Pseudacris, Scaphiopus, Eurycea, Plethodon, Cnemidophorus, Thamnophis, Crotalus.*

521 **Collins, J. T.** 1979. Géographic distribution: *Gastrophryne olivacea.* Herpetol. Rev. 10(3):101.—Distribution. *Gastrophryne.*

522 **Collins, J. T., and J. L. Knight.** 1980. *Crotalus horridus.* Cat. Am. Amphib. Reptiles 253.1-253.2.—Range map and literature review. Bibliographies, Distribution. *Crotalus.*

Collins, J. T., *see* 1196

523 **Collins, R. F., and C. C. Carpenter.** 1968. Organ position-ventral scale relationship in the water moccasin (*Agkistrodon piscivorus leucostoma*), with notes on food habits and distribution. Proc. Okla.

Acad. Sci. 49:15-18.—Morphology, Natural History. *Rana, Nerodia, Agkistrodon.*

524 **Conant, R.** 1949. Two new races of *Natrix erythrogaster.* Copeia 1949(1):1-15.—Taxonomy. *Nerodia.*

525 **Conant, R.** 1958. A field guide to reptiles and amphibians of eastern North America. Houghton Mifflin Co., Boston. 366 pp.—Distribution, Ecology, Natural History, Taxonomy. All Amphibia, All Reptilia.

526 **Conant, R.** 1960. The queen snake, *Natrix septemvittata,* in the Interior Highlands of Arkansas and Missouri, with comments upon similar disjunct distributions. Proc. Acad. Nat. Sci. Phila. 112(2):25-40.—Distribution. *Eurycea, Plethodon, Regina.*

527 **Conant, R.** 1963. Evidence for the specific status of the water snake *Natrix fasciata.* Am. Mus. Novit. (2122):1-38.—Distribution, Taxonomy. *Nerodia.*

528 **Conant, R.** 1971. Reptiles of Oklahoma: By R. G. Webb. Copeia 1971(3):580-581.—A book review. General Reference, Popular.

529 **Conant, R.** 1975. A field guide to reptiles and amphibians of eastern and central North America. Houghton Mifflin Co., Boston. 429 pp.—Distribution, Ecology, Natural History, Taxonomy. All Amphibia, All Reptilia.

530 **Conant, R.** 1978. Distributional patterns of North American snakes. Some examples of the effects of Pleistocene glaciation and subsequent climatic changes. Bull. Md. Herpetol. Soc. 14(4):241-259.—Distribution, Ecology, Paleontology. *Pseudacris, Terrapene, Kinosternon, Heterodon, Nerodia, Thamnophis, Tropidoclonion, Agkistrodon, Crotalus, Sistrurus.*

531 **Conant, R., and C. J. Goin.** 1948. A new subspecies of soft-shelled turtle from the central United States, with comments on the application of the name *Amyda.* Occas. Pap. Mus. Zool. Univ. Mich. (510): 1-19.—Distribution, Taxonomy. *Trionyx.*

Conant, R., *see* 697, 698, 699

Cook, H. J., *see* 745

532 **Cooper, D.** 1927. A journal kept by Douglas Cooper of an expedition by a company of Chickasaw in quest of Comanche Indians. Chron. Okla. 5(4):381-421.—Historical. *Crotalus.*

533 **Cooper, G. M.** 1954. A study of the vertebrate fauna on a small island in Lake Texoma, Oklahoma. M.S. Thesis, Univ. Okla., Norman. 36 pp.—Distribution, Natural History. *Bufo, Acris, Hyla, Rana, Chelydra, Graptemys, Pseudemys, Trachemys, Terrapene, Kinosternon, Trionyx, Sceloporus, Eumeces, Scincella, Cnemidophorus, Coluber, Elaphe, Heterodon, Lampropeltis, Masticophis, Nerodia, Opheodrys, Thamnophis, Agkistrodon, Crotalus.*

534 **Cope, E. D.** 1869. A review of the species of the Plethodontidae and Desmognathidae. Proc. Acad. Nat. Sci. Phila. 21:93-118.—Distribution, Taxonomy. *Eurycea.*

535 **Cope, E. D.** 1889. The Batrachia of North America. Bull. U.S. Natl. Mus. (45):1-525.—Distribution. *Bufo, Acris, Rana, Ambystoma.*

536 **Cope, E. D.** 1893. On a collection of Batrachia and Reptilia from southwest Missouri. Proc. Acad. Nat. Sci. Phila. 45:383-385.—Northeastern region of Oklahoma may have been included. Distribution. *Typhlotriton, Crotaphytus, Sceloporus.*

537 **Cope, E. D.** 1894. On the Batrachia and Reptilia of the Plains at latitude 36'36'. Proc. Acad. Nat. Sci. Phila. 45:386-387.—Distribution, Historical. *Bufo, Pseudacris, Ambystoma, Chelydra, Terrapene, Kinosternon, Ophisaurus, Crotaphytus, Holbrookia, Phrynosoma, Cnemidophorus, Coluber, Diadophis, Heterodon, Lampropeltis, Masticophis, Pituophis, Thamnophis, Tropidoclonion, Crotalus, Sistrurus.*

538 **Cope, E. D.** 1900. The crocodilians, lizards, and snakes of North America. Rep. U.S. Natl. Mus. 1898. Pages 153-1270.—Distribution, Historical, Taxonomy. *Eumeces, Heterodon, Lampropeltis, Storeria, Crotalus.*

539 **Cosgrove, G. E., D. E. Deakins, and J. T. Self.** 1979. Reptilian pentastomiasis. *In* G. L. Hoff (Ed.). Symp. Vol. Wildl. Desease Assoc. Univ. Iowa Press, Iowa City.—Parasitology. *Crotalus.*

540 **Cothran, M. L.** 1978. Effect of melatonin on thermal selection by *Crotaphytus collaris* (Squamata, Iguanidae). M.S. Thesis, Univ. Okla., Norman. 28 pp.—Behavior, Physiology. *Crotaphytus.*

541 **Cothran, M. L., and V. H. Hutchison.** 1979. Effect of melatonin on thermal selection by *Crotaphytus collaris* (Squamata: Iguanidae). Comp. Biochem. Physiol. 63(A):461-466.—Physiology. *Crotaphytus.*

Cox, R., *see* 490

542 **Cragin, F. W.** 1894. Herpetological notes from Kansas and Texas. Colo. Coll. Ann. Publ. 5:37-39.—Distribution, Historical. *Terrapene.*

543 **Crawford, L. E.** 1948. Drama in the woods. Okla. Game Fish News 4(5):11.—Article about opossum killed by diamondback rattlesnake. Popular. *Crotalus.*

544 **Crittendon, W. F.** 1950. A comparative study of the utilization of yolk in some amphibian embryos. M.S. Thesis, Univ. Okla., Norman. 56 pp.—Development, Morphology. *Scaphiopus, Rana.*

545 **Cross, H.** 1917. Animal and plant life in Oklahoma. Okla. Geol. Surv. Circ. (6):1-68.—Distribution. *Bufo, Acris, Hyla, Scaphiopus, Rana, Ambystoma, Chelydra, Macroclemys, Terrapene, Kinosternon, Sternotherus, Trionyx, Ophisaurus, Holbrookia, Phrynosoma, Sceloporus, Eumeces, Carphophis, Coluber, Diadophis, Elaphe, Heterodon, Lampropeltis, Masticophis, Nerodia, Opheodrys, Pituophis, Leptotyphlops, Agkistrodon, Crotalus, Sistrurus, Alligator.*

Cross, J. C., *see* 1255

546 **Crouse, G.** 1966. Panhandle pot luck. Outdoor Okla. 22(3):14-16.—Recipes for rattlesnake included. Popular. *Crotalus.*

547 **Crow, C. L.** 1959. Okla. Wildl. 15(6):15.—Photograph of freshly killed timber rattlesnake. Incorrectly printed as issue 4. Popular. *Crotalus.*

548 **Crozier, B., and J. T. Self.** 1941. A new host record of the trematode: *Neorenifer serpentis.* Proc. Okla. Acad. Sci. 21:31.—Parasitology. *Heterodon.*

549 **Cuellar, H. S.** 1971. Levels of genetic compatibility of *Rana areolata* with southwestern members of the *Rana pipiens* complex (Anura: Ranidae). Evolution 25(2):399-409.—Genetics. *Rana.*

550 **Curd, M. R.** 1950. The salamander *Amphiuma tridactylum* in Oklahoma. Copeia 1950(4):324.—Distribution. *Amphiuma, Farancia, Nerodia, Agkistrodon.*

551 **Daly, E.** 1969. A new procolophoniod reptile from the Lower Permian of Oklahoma. J. Paleontol. 43(3):676-687.—General Reference, Paleontology.

552 **Daly, E.** 1973. A Lower Permian vertebrate fauna from southern Oklahoma. J. Paleontol. 47(3):562-589.—General Reference, Paleontology.

David, P. R., *see* 869

Davis, D. D., *see* 1271

553 **Davis, D. M.** 1969. A study of variation in North American lizards of the *fasciatus* group of the genus *Eumeces* (Scincidae). Ph.D. Diss., Duke Univ., Durham. 171 pp.—Distribution, Evolution, Morphology. *Eumeces.*

554 **Davis, G. D.** 1960. A simple method for opening the turtle plastron. Proc. Okla. Acad. Sci. 41:82-77.—Techniques. *Trachemys.*

555 **Davis, H.** 1946. Okla. Game Fish News 2(8):cover photo.—Photograph of a roadrunner attacking a watersnake. Lore, Popular. *Nerodia.*

Dawson, D., *see* 153

556 **Day, D.** 1962. Pet turtles. Okla. Wildl. 18(5):21.—Popular account of having turtles for pets. Popular. *Terrapene.*

557 **Deakins, D. E.** 1973. Pentastome pathology in captive reptiles. Ph.D. Diss., Univ. Okla., Norman. 52 pp.—Parasitology. *Nerodia, Thamnophis.*

Deakins, D. E., *see* 539

Dean, R. H., *see* 574

558 **Decker, C. E.** 1930. Age of the toads in the A. H. Holloman Sand Pit at Frederick, Oklahoma. Proc. Okla. Acad. Sci. 10:82-84.—Live toads found in a Pleistocene deposit. Paleontology. *Bufo, Scaphiopus.*

559 **deRicqles, A., and J. R. Bolt.** 1983. Jaw growth and tooth replacement in *Captorhinus aguti* (Reptilia: Captorhinomorpha): A morphological and histological analysis. J. Vert. Paleontol. 3(1):7-24.—General Reference, Paleontology.

DeMar, R., *see* 222

Derrick, G. E., *see* 850

Dessauer, H. C., *see* 673

560 **Diener, R. A.** 1957. An ecological study of the plain-bellied water snake. Herpetologica 13(3):203-211.—Ecology. *Nerodia.*

561 **Diener, R. A.** 1957. A western hognose snake eats a collared lizard. Herpetologica 13(2):122.—Natural History. *Crotaphytus, Heterodon.*

562 **Diener, R. A.** 1961. Notes on a bite of the broad-banded copperhead, *Ancistrodon contortrix laticinctus* Gloyd and Conant. Herpetologica 17(2):143-144.—Behavior. *Agkistrodon.*

563 **DiSilvestro, R.** 1979. The heat is on. Outdoor Okla. 35(7):14-16.—Popular. *Terrapene, Trionyx, Crotaphytus, Alligator.*

564 **DiSilvestro, R.** 1980. Autumn animal changes. Outdoor Okla. 36(8):34-41.—Included the dormancy of frogs, lizards, and rattlesnakes.

565 **DiSilvestro, R.** 1980. Nature's children. Outdoor Okla. 36(4):8-11.—Included foraging habits of tadpoles and turtles. Popular. *Chelydra.*

566 **DiSilvestro, R.** 1981. Death. Outdoor Okla. 37(1):39-43.—Article on predation. Popular. *Lampropeltis.*

567 **DiSilvestro, R.** 1981. Strange critters. Outdoor Okla. 37(2):40-45.—Reviews atypical colors. Popular. *Rana, Agkistrodon.*

568 **DiSilvestro, R.** 1982. Frogs and toads. Outdoor Okla. 27(3):15-19.—Popular. *Bufo, Rana.*

569 **DiSilvestro, R.** 1982. Wild vision. Outdoor Okla. 38(5):25-28.—Popular. *Rana.*

570 **DiSilvestro, R.** 1983. Salamanders. Outdoor Okla. 39(4):38-43.—Natural History, Popular. *Ambystoma, Amphiuma, Desmognathus, Eurycea, Plethodon, Typhlotriton, Notophthalmus.*

571 **Ditmars, R. L.** 1930. The poisonous serpents of the New World. Bull. N.Y. Zool. Soc. 33:79-132.—Distribution. *Agkistrodon, Crotalus, Sistrurus.*

572 **Dixon, J. R.** 1959. Geographic variation and distribution of the long-tailed group of the glossy snake, *Arizona elegans* Kennicott. Southwest. Nat. 4(1):20-29.—Distribution, Morphology. *Arizona.*

573 **Dixon, J. R.** 1965. A taxonomic reevaluation of the night snake *Hypsiglena ochrorhyncha* and relatives. Southwest. Nat. 10(2):125-131.—Taxonomy. *Hipsiglena.*

574 **Dixon, J. R., and R. H. Dean.** 1986. Status of the southern population of the night snake (*Hypsiglena*: Colubridae) exclusive of California and Baja California. Southwest. Nat. 31(3):307-318.—Distribution, Taxonomy. *Hypsiglena.*

575 **Dixon, J. R., and R. R. Fleet.** 1976. *Arizona, A. elegans.* Cat. Am. Amphib. Reptiles 179.1-179.4.—Range map and literature review. Bibliographies, Distribution. *Arizona.*

Dixon, J. R., *see* 641

576 **Dodd, C. K., Jr.** 1980. Notes on the feeding behavior of the Oklahoma salamander, *Eurycea tynerensis* (Plethodontidae). Southwest. Nat. 25(1):111-113.—Behavior, Ecology. *Eurycea.*

577 **Dodd, C. K., Jr.** 1982. Correction: *Typhlotriton spelaeus* not *Eurycea tynerensis.* Southwest. Nat. 27(3):367-368.—Taxonomy. *Eurycea, Typhlotriton.*

578 **Dodd, C. K., Jr.** 1986. Endangered and threatened wildlife and plants; a review of vertebrate wildlife for listing as endangered or threatened species. Fed. Regist. 50(181):37958-37967.—Management. *Macroclemys, Eurycea, Plethodon, Lampropeltis.*

579 **Dodd, C. K., Jr., and M. T. Christensen.** 1977. Percent assimilation and feeding behavior of the Oklahoma salamander, *Eurycea tynerensis.* A.B.S. Bull. 24(2):47.—Behavior, Physiology. *Eurycea.*

Dodd, C. K., Jr., *see* 424

580 **Dodson, P., R. T. Bakker, and A. K. Behrensmeyer.** 1983. Paleoecology of the dinosaur-bearing Morrison Formation. Res. Rep. Natl. Geogr. Soc. 15:145-156.—Faunal list included amphibians and reptiles from Oklahoma. General Reference, Paleontology.

581 **Douglas, K.** 1978. The animal kingdom in glass cages. Univ. Okla. Mag.—General Reference, Historical, Popular.

Dowell, V. E., *see* 367

582 **Dowling, H. G.** 1951. A taxonomic study of the ratsnakes, genus *Elaphe* Fitzinger. I. The status of the name *Scotophis laetus* Baird and Girard (1853). Copeia 1951(1):39-44.—Distribution, Taxonomy. *Elaphe.*

583 **Dowling, H. G.** 1956. Geographic relations of Ozarkian amphibians and reptiles. Southwest. Nat. 1(4):174-189.—Distribution. *Ambystoma, Desmognathus, Eurycea, Plethodon, Typhlotriton.*

584 **Dowling, H. G.** 1959. Classification of the serpentes: A critical review. Copeia 1959(1):38-52.—Taxonomy. *Leptotyphlops.*

585 **Drass, R. R., T. G. Baugh, and P. Flynn.** 1987. The Heerwald Site and early Plains Village adaptations on the southern Plains. N.A. Archeol. 8(2):151-167.—Turtle shells were found. *Chelydra, Trachemys, Terrapene, Trionyx.*

586 **Drass, R. R., and P. Flynn.** 1986. Temporal and geographic variations in subsistence practices for Plains Villagers in the southern

Plains. 44th Southern Plains Conference, Denver.—Turtle shells were found. Archeology. *Chelydra, Trachemys, Terrapene.*

Drass, R. R., *see* 402

587 **Duellman, W. E., and R. C. Zweifel.** 1962. A synopsis of the lizards of the *sexlineatus* group (Genus *Cnemidophorus*). Bull. Am. Mus. Nat. Hist. 123(3):159-210.—Distribution, Morphology. *Cnemidophorus.*

588 **Duncan, R., and R. Highton.** 1979. Genetic relationships of the eastern large *Plethodon* of the Ouachita Mountains. Copeia 1979(1):95-110.—Distribution, Genetics. *Plethodon.*

589 **Dundee, H. A.** 1947. Notes on salamanders collected in Oklahoma. Copeia 1947(2):117-120.—Distribution, Natural History. *Ambystoma, Eurycea, Plethodon, Typhlotriton.*

590 **Dundee, H. A.** 1948. *Storeria occipitomaculata* in Oklahoma. Copeia 1948(3):216.—Distribution. *Storeria.*

591 **Dundee, H. A.** 1950. Additional records of *Hypsiglena* from Oklahoma with notes on the behavior and eggs. Herpetologica 6(2):28-30.—Behavior, Natural History. *Hypsiglena.*

592 **Dundee, H. A.** 1950. An improved method for collecting living lizards and frogs. Herpetologica 6(3):78-79.—Techniques. *Crotaphytus.*

593 **Dundee, H. A.** 1950. *Kinosternon subrubrum hippocrepis* (Gray) in Oklahoma. Herpetologica 6(5):138-139.—Distribution. *Kinosternon.*

594 **Dundee, H. A.** 1950. Notes on the type locality of *Eurycea multiplicata* (Cope). Herpetologica 6(2):27-28.—Distribution. *Eurycea.*

595 **Dundee, H. A.** 1950. The type locality for *Elaphe l. laeta* (B. & G.). Herpetologica 6(2):55.—Distribution. *Elaphe.*

596 **Dundee, H. A.** 1965. *Eurycea multiplicata.* Cat. Am. Amphib. Reptiles 21.1-21.2.—Range map and literature review. Bibliographies, Distribution. *Eurycea.*

597 **Dundee, H. A.** 1965. *Eurycea tynerensis.* Cat. Am. Amphib. Reptiles 22.1-22.2.—Range map and literature review. Bibliographies, Distribution. *Eurycea.*

598 **Dundee, H. A., and A. N. Bragg.** 1945. Extensions of known salientian ranges in Oklahoma. Proc. Okla. Acad. Sci. 26:18-19.—Distribution. *Bufo, Hyla, Pseudacris, Scaphiopus, Rana.*

599 **Dundee, H. A., and A. N. Bragg.** 1950. A key to the adult salamanders of Oklahoma, including the forms likely to occur in the

state. Pages 18-19 *in* Researches on the Amphibia of Oklahoma. Univ. Okla. Press, Norman.—Taxonomy, Morphology. All Caudata.

600 **Dundee, H. A., and W. L. Burger, Jr.** 1948. A denning aggregation of the western cottonmouth. Nat. Hist. Misc. (Chic.) (21):1-2.—Ecology, Natural History. *Agkistrodon.*

Dundee, H. A., *see* 368, 369, 386, 691

601 **Dunham, J. M.** 1950. Some helminths of *Natrix rhombifera rhombifera* and *Natrix erythrogaster transversa* in Payne County. M.S. Thesis, Okla. A.M. Coll., Stillwater. 42 pp.—Parasitology. *Nerodia.*

602 **Dunn, E. R.** 1926. The salamanders of the family Plethodontidae. Smith Coll., Northampton, Mass. 441 pp.—Distribution, Taxonomy. *Desmognathus, Eurycea.*

603 **Dunn, E. R.** 1932. The status of *Tropidoclonion lineatum.* Proc. Biol. Soc. Wash. (45):195-198.—Distribution, Morphology, Taxonomy. *Tropidoclonion.*

604 **Dunn, E. R.** 1938. Notes on frogs of the genus *Acris.* Proc. Acad. Nat. Sci. Phila. 90:153-154.—Distribution, Morphology, Taxonomy. *Acris.*

605 **Dunn, E. R.** 1940. The races of *Ambystoma tigrinum.* Copeia 1940(3):154-162.—Distribution, Taxonomy. *Ambystoma.*

606 **Dunn, E. R., and A. A. Heinze.** 1933. A new salamander from the Ouachita Mountains. Copeia 1933(3):121-122.—Distribution, Morphology, Taxonomy. *Plethodon.*

Dunn, E. R., *see* 658

Duvall, D., *see* 482, 1361

607 **Dyer, W. G., and R. A. Brandon.** 1973. Helminths from salamanders in Oklahoma and Missouri. Herpetologica 29(4):371-373.—Parasitology. *Ambystoma.*

Edds, D. R., *see* 518

608 **Edgren, R. A.** 1952. Biogeographical and behavioral considerations of the snakes of the genus *Heterodon.* Ph.D. Diss., Northwestern Univ., Evanston, IL. 232 pp.—Distribution, Natural History, Taxonomy. *Heterodon.*

609 **Edgren, R. A.** 1952. A synopsis of the snakes of the genus *Heterodon,* with the diagnosis of a new race of *Heterodon nasicus* Baird and

Girard. Nat. Hist. Misc. (Chic.) (112):1-4.—Distribution, Taxonomy. *Heterodon.*

610 **Edgren, R. A.** 1957. Melanism in hog-nosed snakes. Herpetologica 13(2):131-135.—Morphology. *Heterodon.*

611 **Edgren, R. A.** 1961. A simplified method for analysis of clines; geographic variation in the hognose snake *Heterodon platyrhinos* Latreille. Copeia 1961(2):125-132.—Taxonomy. *Heterodon.*

Edwards, S. R., *see* 1196

612 **Ellis, R. J.** 1951. Food habits and control of coyotes in northcentral Oklahoma. M.S. Thesis, Okla. A.M. Coll., Stillwater. 45 pp.—Unidentified traces of snakes and lizards were found in scats. General Reference, Natural History.

613 **Ellis, R. J., and S. D. Schemnitz.** 1957. Some foods used by coyotes and bobcats in Cimarron County, Oklahoma, 1954-56. Proc. Okla. Acad. Sci. 38:180-185.—Natural History. *Holbrookia.*

Endicott, C. G., *see* 3

614 **Engbretson, G.** 1974. Amphibians and reptiles of Oklahoma. Pages 103-188 *in* P. G. Risser (Ed.). Field guide to Oklahoma. Okla. Biol. Surv. 160 pp.—Checklist of Oklahoma herpetofauna with notes on their occurrence in the state. Distribution, Natural History. All Amphibia, All Reptilia.

615 **England, M.** 1983. Red-winged blackbird defends nest from black rat snake. Bull. Okla. Herp. Soc. 8(4):103-104.—Natural History. *Elaphe.*

England, M., *see* 1443

616 **Ernst, C. H.** 1971. *Chrysemys picta.* Cat. Am. Amphib. Reptiles 106.1-106.4.—Range map and literature review. Bibliographies, Distribution. *Chrysemys.*

617 **Ernst, C. H., and R. W. Barbour.** 1972. Turtles of the United States. Univ. Kentucky Press, Lexington. 347 pp.—Behavior, Distribution, Natural History, Taxonomy. All Testudines.

618 **Erskine, D. J., and V. H. Hutchison.** 1981. Melatonin and behavioral thermoregulation in the turtle, *Terrapene carolina triunguis.* Physiol. Behav. 26(6):991-994.—Physiology. *Terrapene.*

Erwin, R. L., *see* 1247

619 **Etheridge, R.** 1960. The slender glass lizard, *Ophisaurus attenuatus,* from the Pleistocene (Illinoian glacial) of Oklahoma. Copeia 1960(1): 46-47.—Paleontology. *Ophisaurus.*

620 **Etheridge, R.** 1961. Late Coenozoic glass lizards (*Ophisaurus*) from the southern Great Plains. Herpetologica 17(3):179-186.—Distribution, Paleontology. *Ophisaurus.*

621 **Everhart, B. A.** 1956. A survey and comparison of the helminth parasites of the red-eared turtle, *Pseudemys scripta elegans* (Wied, 1838) Carr, 1938, in Texas and Oklahoma. M.S. Thesis, Okla. A.M. Coll., Stillwater. 80 pp.—Parasitology. *Trachemys.*

622 **Everhart, B. A.** 1957. Notes on the helminths of *Pseudemys scripta elegans* (Wied, 1838) in areas of Texas and Oklahoma. Proc. Okla. Acad. Sci. 38:38-43.—Parasitology. *Trachemys.*

623 **Fahey, K. M.** 1980. A taxonomic study of the cooter turtles, *Pseudemys floridana* (LeConte) and *Pseudemys concinna* (LeConte) in the lower Red River, Atchafalaya River, and Mississippi River basins. Tulane Stud. Zool. Bot. 22(2):49-66.—Distribution, Taxonomy. *Pseudemys.*

Falk, C. R., *see* 1067

624 **Faxon, W.** 1915. Relics of Peale's Museum. Bull. Mus. Comp. Zool. 59(3):117-148.—Distribution, General Reference, Historical.

Fellers, G. M., *see* 424

625 **Ferguson, H. N.** 1932. Spermatogenesis of *Natrix sipedon sipedon* with special reference to the sex chromosome. M.S. Thesis, Univ. Okla., Norman. 19 pp.—Development, Distribution. *Nerodia.*

626 **Fesperman, W.** 1986. Notes on amphibians of northern LeFlore County, Oklahoma. Bull. Okla. Herp. Soc. 11(1-4):40-41.—Distribution, Natural History. *Bufo, Pseudacris, Gastrophryne, Rana, Ambystoma, Eurycea.*

627 **Fesperman, W.** 1987. Geographic distribution: *Tropidoclonion lineatum annectans.* Herpetol. Rev. 18(1):21.—Distribution. *Tropidoclonion.*

628 **Firschein, I. L., and L. S. Miller.** 1951. The ringed salamander, *Ambystoma annulatum,* in Oklahoma. Copeia 1951(1):82.—Distribution. *Ambystoma.*

629 **Fisher, H. T., and A. Richards.** 1950. The annual ovarian cycle of *Acris crepitans* Baird. Pages 129-142 *in* Researches on the Amphibia

of Oklahoma. Univ. Okla. Press, Norman.—Behavior, Ecology, Morphology, Physiology. *Acris.*

Fisher, H. T., *see* 386

630 **Fitch, H. S.** 1940. A biogeographic study of the *ordinoides* artenkreis of garter snakes (Genus *Thamnophis*). Univ. Calif. Publ. Zool. 44(1): 1-150.—Distribution, Taxonomy. *Thamnophis.*

631 **Fitch, H. S.** 1954. Life history and ecology of the five-lined skink, *Eumeces fasciatus.* Univ. Kans. Publ. Mus. Nat. Hist. 8(1):1-156.—Distribution, Ecology. *Eumeces.*

632 **Fitch, H. S.** 1955. Habits and adaptations of the Great Plains skink (*Eumeces obsoletus*). Ecol. Monogr. 25(1):59-83.—Distribution. *Eumeces.*

633 **Fitch, H. S.** 1960. Autecology of the copperhead. Univ. Kans. Publ. Mus. Nat. Hist. 13(4):85-288.—Distribution, Morphology. *Agkistrodon.*

634 **Fitch, H. S.** 1965. An ecological study of the garter snake, *Thamnophis sirtalis.* Univ. Kans. Publ. Mus. Nat. Hist. 15(10):493-504.—Distribution, Ecology. *Thamnophis.*

635 **Fitch, H. S.** 1978. Sexual size differences in the genus *Sceloporus.* Univ. Kans. Sci. Bull. 51(13):441-461.—Morphology. *Sceloporus.*

636 **Fitch, H. S.** 1980. Reproductive strategies of reptiles. Pages 25-32 *in* J. B. Murphy and J. T. Collins (Eds.). Reproductive biology and diseases of captive reptiles. Soc. Study Amphib. Reptiles Contrib. Herpetol. (1):33-48—Clutch size data from an Oklahoma location is included. Natural History. *Scincella.*

637 **Fitch, H. S.** 1980. *Thamnophis sirtalis.* Cat. Am. Amphib. Reptiles 270.1-270.4.—Range map and literature review. Bibliographies, Distribution. *Thamnophis.*

638 **Fitch, H. S.** 1983. *Thamnophis elegans.* Cat. Am. Amphib. Reptiles 320.1-320.4.—Range map and literature review. Bibliographies, Distribution. *Thamnophis.*

639 **Fitch, H. S., and H. W. Greene.** 1965. Breeding cycle in the ground skink, *Lygosoma laterale.* Univ. Kans. Publ. Mus. Nat. Hist. 15(11): 565-575.—Natural History. *Scincella.*

640 **Fitch, H. S., and T. P. Maslin.** 1961. Occurrence of the garter snake, *Thamnophis sirtalis* in the Great Plains and Rocky Mountains. Univ.

Kans. Publ. Mus. Nat. Hist. 13(5):289-308.—Distribution, Taxonomy. *Thamnophis.*

Fiuso, L. F., *see* 1363

641 **Fleet, R. R., and J. R. Dixon.** 1971. Geographic variation within the long-tailed group of the glossy snake, *Arizona elegans* Kennicott. Herpetologica 27(3):295-302.—Distribution, Taxonomy. *Arizona.*

Fleet, R. R., *see* 575

642 **Flynn, P.** 1984. An analysis of the 1973 test excavation at the Zimms Site (34rm72). Pages 215-290 *in* Archaeology of the mixed grass prairie. Phase 1: Quartermaster Creek. Okla. Archeol. Res. Surv. Rep. (20):1-313.—Archeology. *Terrapene.*

Flynn, P., *see* 585, 586, 672

643 **Fong, D. W.** 1980. Ecology and behavior of the Great Plains ground snake, *Sonora episcopa episcopa,* in Oklahoma. M.S. Thesis, Univ. Okla., Norman. 71 pp.—Behavior, Ecology. *Sonora.*

644 **Force, E. R.** 1925. Notes on reptiles and amphibians of Okmulgee County, Oklahoma. Copeia 1925(141):25-27.—Distribution, Natural History. *Bufo, Acris, Hyla, Rana, Chelydra, Terrapene, Kinosternon, Ophisaurus, Crotaphytus, Sceloporus, Eumeces, Cnemidophorus, Coluber, Diadophis, Elaphe, Heterodon, Lampropeltis, Masticophis, Nerodia, Opheodrys, Pituophis, Regina, Storeria, Thamnophis, Tropidoclonion.*

645 **Force, E. R.** 1925. Notes on reptiles and amphibians of Okmulgee County, Oklahoma. Proc. Okla. Acad. Sci. 5:80-83.—Distribution. *Bufo, Acris, Hyla, Rana, Chelydra, Terrapene, Kinosternon, Ophisaurus, Crotaphytus, Sceloporus, Eumeces, Cnemidophorus, Coluber, Diadophis, Elaphe, Heterodon, Lampropeltis, Masticophis, Nerodia, Opheodrys, Pituophis, Regina, Storeria, Thamnophis, Tropidoclonion.*

646 **Force, E. R.** 1927. The bird collection of Okmulgee High School. Proc. Okla. Acad. Sci. 7:57-71.—Snake skins were found in nests. General Reference, Natural History.

647 **Force, E. R.** 1928. A preliminary checklist of amphibians and reptiles of Tulsa County, Oklahoma. Proc. Okla. Acad. Sci. 8:78-79.— Distribution. *Bufo, Acris, Ambystoma, Chelydra, Trachemys, Terrapene, Kinosternon, Ophisaurus, Crotaphytus, Phrynosoma, Sceloporus, Eumeces, Scincella, Cnemidophorus, Carphophis, Coluber, Diadophis, Elaphe, Heterodon, Lampropeltis, Masticophis, Nerodia, Opheodrys, Pituophis, Regina, Storeria, Tantilla, Thamnophis, Tropidoclonion, Virginia, Leptotyphlops, Agkistrodon.*

648 **Force, E. R.** 1930. The amphibians and reptiles of Tulsa County, Oklahoma, and vicinity. Copeia 1930(2):25-39.—Distribution, Ecology, Natural History. *Bufo, Acris, Hyla, Pseudacris, Gastrophryne, Rana, Ambystoma, Necturus, Chelydra, Trachemys, Terrapene, Kinosternon, Sternotherus, Trionyx, Ophisaurus, Crotaphytus, Phrynosoma, Sceloporus, Eumeces, Scincella, Cnemidophorus, Carphophis, Cemophora, Coluber, Diadophis, Elaphe, Heterodon, Lampropeltis, Masticophis, Nerodia, Opheodrys, Pituophis, Regina, Sonora, Storeria, Tantilla, Thamnophis, Tropidoclonion, Virginia, Leptotyphlops, Agkistrodon, Crotalus.*

649 **Force, E. R.** 1931. Habits and birth of young of the lined snake, *Tropidoclonion lineatum* (Hallowell). Copeia 1931(2):51-53.—Natural History. *Tropidoclonion.*

650 **Force, E. R.** 1934. A local study of the opisthoglyph snake *Tantilla gracilis* Baird and Girard. Pap. Mich. Acad. Sci. Arts Lett. (20):645-659.—Behavior, Morphology, Natural History. *Tantilla.*

651 **Force, E. R.** 1935. Notes on the blind snake, *Leptotyphlops dulcis* (Baird and Girard) in northeastern Oklahoma. Proc. Okla. Acad. Sci. 16:24-26.—Natural History. *Sonora, Tantilla, Leptotyphlops.*

652 **Force, E. R.** 1935. The relation of the knobbed anal keels to age and sex in the lined snake *Tropidoclonion lineatum* (Hallowell). Pap. Mich. Acad. Sci. Arts Lett. (21):613-617.—Morphology, Natural History. *Tropidoclonion.*

653 **Force, E. R., and N. L. Schmerchel.** 1933. Sexual differentiation in the tail length of the lined snake *Tropidoclonion lineatum* (Hallowell). Biologist 14(2):70-72.—Morphology. *Tropidoclonion.*

Force, E. R., *see* 205

654 **Ford, N. B.** 1976. Pheromone trailing behavior in three species of garter snake (*Thamnophis*). M.S. Thesis, Univ. Okla., Norman. 32 pp.—Behavior. *Thamnophis.*

655 **Ford, N. B.** 1978. Evidence for species specificity of pheromone trails in two sympatric gartersnakes, *Thamnophis.* Herpetol. Rev. 9(1):10.—An abstract. Behavior. *Thamnophis.*

656 **Ford, N. B.** 1981. Seasonality of pheromone trailing behavior in two garter snakes, *Thamnophis* (Colubridae). Southwest. Nat. 26(4):385-388.—Behavior. *Thamnophis.*

657 **Foreman, G.** 1934. Survey of a wagon road from Fort Smith to the Colorado River. Chron. Okla. 12(1):74-96.—Snakes were included. Historical. *Rana, Trachemys.*

Fouquette, M. J., Jr., *see* 968

658 **Fowler, H. W., and E. R. Dunn.** 1917. Notes on salamanders. Proc. Acad. Nat. Sci. Phila. Jan. 28 pp.—Distribution. *Ambystoma.*

659 **Fox, R. C.** 1962. Two new pelycosaurs from the Lower Permian of Oklahoma. Univ. Kans. Publ. Mus. Nat. Hist. 12(6):297-307.—General Reference, Paleontology.

660 **Fox, R. C.** 1964. The adductor muscle of the jaw in some primitive reptiles. Univ. Kans. Publ. Mus. Nat. Hist. 12(15):657-680.—Oklahoma fossils used. General Reference, Paleontology.

661 **Fox, R. C., and M. C. Bowman.** 1966. Osteology and relationships of *Captorhinus aguti* (Cope) (Reptilia: Captorhinomorpha). Univ. Kans. Paleontol. Contrib. Vert. 79 pp.—General Reference, Paleontology.

662 **Fraser, J. C.** 1985. Southeast Oklahoma at it's best. Kans. Herpetol. Soc. Newsl. (59):7-8.—Account of a collecting trip to Oklahoma. Distribution, Popular. *Plethodon, Sceloporus, Eumeces, Scincella, Carphophis, Coluber, Diadophis, Lampropeltis, Masticophis, Sonora, Tantilla.*

663 **Freeman, B.** 1929. Variations in the lepidosis of some Oklahoma snakes. M.S. Thesis, Univ. Okla., Norman. 136 pp.—Morphology, Techniques. *Diadophis, Nerodia, Thamnophis, Tropidoclonion.*

Freeman, B., *see* 1159

664 **Freeman, S. L., R. L. Lardie, J. Pigg, and J. H. Black.** 1987. Distribution records for some snakes in Oklahoma. Bull. Okla. Herp. Soc. 12(1-4):13-15.—Distribution. *Lampropeltis, Nerodia, Pituophis.*

665 **Frost, D. R.** 1983. *Sonora semiannulata.* Cat. Am. Amphib. Reptiles 333.1-333.4.—Range map and literature review. Bibliographies, Distribution. *Sonora.*

666 **Frost, D. R.** 1985. Amphibian species of the world: A taxonomic and geographic reference. Allen Press, Inc., Lawrence. 732 pp.—Taxonomy. All Amphibians.

Frost, J. S., *see* 779

667 **Fulhage, I.** 1954. The occurrence of *Cotylaspis insignis* Leidy (Aspidogastridae) in clams and turtles of Lake Texoma. Proc. Okla. Acad. Sci. 35:67-68.—Flukes were found in softshell turtles. Parasitology. *Trionyx.*

668 **Funk, R. S., and J. K. Tucker.** 1978. Variation in a large brood of lined snakes, *Tropidoclonion lineatum* (Reptilia, Serpentes, Colubridae). J. Herpetol. 12(1):115-117.—Morphology. *Tropidoclonion.*

669 **Galbreath, E. C.** 1948. A new extinct emydid turtle from the Lower Pliocene of Oklahoma. Univ. Kans. Publ. Mus. Nat. Hist. 1(16):265-275.—Paleontology. *Chrysemys.*

670 **Galbreath, E. C.** 1948. Pliocene and Pleistocene records of fossil turtles from western Kansas and Oklahoma. Univ. Kans. Publ. Mus. Nat. Hist. 1(17):283-284.—General Reference, Paleontology.

671 **Galm, J. R.** 1978. Archeological investigation at Wister Lake, Le-Flore County, Oklahoma. Univ. Okla. Archeol. Res. Manage. Cent. Res. Ser. (1):1-279.—Emydid turtles were found. Archeology, General Reference.

672 **Galm, J. R., and P. Flynn.** 1978. The cultural sequence at the Scott (34lf-11) and Wann (34lf-27) sites and prehistory of the Wister Valley. Univ. Okla. Archeol. Res. Manage. Cent. Res. Ser. (3):1-350.—Archeology. *Chelydra, Graptemys, Trachemys, Terrapene, Trionyx.*

Galm, J. R., *see* 212

673 **Gartside, D. F., J. S. Rogers, and H. C. Dessauer.** 1977. Speciation with little genetic and morphological differentiation in the ribbon snake *Thamnophis proximus* and *T. sauritus* (Colubridae). Copeia 1977(4):697-705.—Distribution, Genetics. *Thamnophis.*

674 **Gates, C. E., J. Pigg, and J. H. Black.** 1986. Verification of the yellow mud turtle in Pottawatomie and Grant counties, Oklahoma. Bull. Okla. Herp. Soc. 11(1-4):19.—Distribution. *Kinosternon.*

675 **Gehlbach, F. R.** 1956. Annotated records of southwestern amphibians and reptiles. Trans. Kans. Acad. Sci. 59(3):364-372.—Distribution, Natural History. *Sceloporus, Cnemidophorus, Elaphe, Masticophis.*

676 **Gehlbach, F. R.** 1965. Amphibians and reptiles from the Pliocene and Pleistocene of North America: A chronological summary and selected bibliography. Tex. J. Sci. 17(1):56-70.—Paleontology. *Trionyx, Ophisaurus.*

677 **Gehlbach, F. R.** 1967. *Ambystoma tigrinum.* Cat. Am. Amphib. Reptiles 52.1-52.4.—Range map and literature review. Bibliographies, Distribution. *Ambystoma.*

678 Gehlbach, F. R. 1970. Death-feigning and erratic behavior in leptotyphlopid, colubrid, and elapid snakes. Herpetologica 26(1):24-34.—Behavior, Techniques. *Lampropeltis.*

679 Gehlbach, F. R., J. F. Watkins II, and J. C. Kroll. 1971. Pheromone trail-following studies of typhlopid, leptotyphlopid and colubrid snakes. Behaviour 40(3-4):282-294.—Behavior. *Leptotyphlops.*

Gehlbach, F. R., *see* 1469, 1470

680 Gehrmann, W. H., and C. C. Carpenter. 1973. Evidence for a central cholinoceptive component in despotic behavior in the male collared lizard, *Crotaphytus collaris.* Proc. Okla. Acad. Sci. 53:38-40.—Behavior, Physiology. *Crotaphytus.*

Gentry, J. L., Jr., *see* 1272

681 Gillingham, J. C. 1976. Comparative reproductive behavior of the rat snakes of eastern North America, genus *Elaphe.* Ph.D. Diss., Univ. Okla., Norman. 68 pp.—Behavior, Techniques. *Elaphe.*

682 Gillingham, J. C. 1979. Reproductive behavior of the rat snakes of eastern North America, genus *Elaphe.* Copeia 1979(2):319-332.—Behavior. *Elaphe.*

683 Gillingham, J. C. 1980. Communication and combat behavior of the black rat snake (*Elaphe obsoleta*). Herpetologica 36(2):120-127.—Behavior. *Elaphe.*

684 Gillingham, J. C., and R. E. Baker. 1981. Evidence for scavenging behavior in the western diamondback rattlesnake (*Crotalus atrox*). Z. Tierpsychol. 55(3):217-227.—Behavior. *Elaphe.*

685 Gillingham, J. C., and C. C. Carpenter. 1978. Snake hibernation: Construction of and observations on a man-made hibernaculum (Reptilia, Serpentes). J. Herpetol. 12(4):495-498.—Management, Techniques. *Elaphe, Pituophis, Agkistrodon.*

686 Gillingham, J. C., and T. Rush. 1974. Notes on the fishing behavior of water snakes. J. Herpetol. 8(4):384-385.—Behavior. *Nerodia.*

Gillingham, J. C., *see* 479

Gilstrap, R. L., *see* 721

687 Ginn, R. 1979. Poisonous snakes of Oklahoma. Bull. Okla. Herp. Soc. 4(2-3):43-44.—Natural History, Popular. *Agkistrodon, Crotalus, Sistrurus.*

Girard, C., *see* 79, 80

688 **Glass, B. P.** 1949. Records of *Macrochelys temminckii* in Oklahoma. Copeia 1949(2):138-141.—Distribution. *Macroclemys.*

689 **Glass, B. P.** 1951. Age at maturity of neotenic *Ambystoma tigrinum mavortium* Baird. Am. Midl. Nat. 46(2):391-394.—Natural History. *Ambystoma.*

690 **Glass, B. P.** 1975. Mammals, reptiles, and amphibians. Pages 13-21. *in* Rare and endangered vertebrates and plants of Oklahoma. Dept. Agri. Soil Cons. Serv. 44 pp.—Prepared by Oklahoma Committee on rare and endangered species. Management. *Bufo, Hyla, Scaphiopus, Rana, Ambystoma, Amphiuma, Plethodon, Typhlotriton, Siren, Chrysemys, Graptemys, Phrynosoma, Sceloporus, Uta, Cnemidophorus, Farancia, Hypsiglena, Regina, Storeria, Virginia.*

691 **Glass, B. P., and H. A. Dundee.** 1950. *Cnemidophorus tesselatus* (Say) in Oklahoma. Herpetologica 6(2):30.—Distribution, Natural History. *Cnemidophorus.*

692 **Glass, B. P., and R. L. Lardie.** 1976. Another Oklahoma record of the northern scarlet snake. Bull. Okla. Herp. Soc. 1(3):34.—Distribution. *Cemophora.*

Glass, B. P., *see* 946

693 **Gloyd, H. K.** 1935. The subspecies of *Sistrurus miliarius.* Occas. Pap. Mus. Zool. Univ. Mich. (322):1-7.—Distribution, Taxonomy. *Sistrurus.*

694 **Gloyd, H. K.** 1940. The rattlesnakes, genera *Sistrurus* and *Crotalus.* Spec. Publ. Chic. Acad. Sci. 4(1):1-266.—Bibliographies, Distribution, Taxonomy. *Crotalus, Sistrurus.*

695 **Gloyd, H. K.** 1955. A review of the massasaugas, *Sistrurus catenatus,* of the southwestern United States (Serpentes: Crotalidae). Bull. Chic. Acad. Sci. 10(6):83-98.—Distribution, Morphology, Taxonomy. *Crotalus, Sistrurus.*

696 **Gloyd, H. K.** 1969. Two additional subspecies of North American crotalid snakes, genus *Agkistrodon.* Proc. Biol. Soc. Wash. (82):219-232.—Distribution, Taxonomy. *Agkistrodon.*

697 **Gloyd, H. K., and R. Conant.** 1934. The broad-banded copperhead: A new subspecies of *Agkistrodon mokasen.* Occas. Pap. Mus. Zool. Univ. Mich. (283):1-5.—Distribution, Morphology, Taxonomy. *Agkistrodon.*

698 **Gloyd, H. K., and R. Conant.** 1938. The subspecies of the copperhead, *Agkistrodon mokasen* Beauvois. Bull. Chic. Acad. Sci. 5(7): 163-166.—Distribution, Morphology, Taxonomy. *Agkistrodon.*

699 **Gloyd, H. K., and R. Conant.** 1943. A synopsis of the American forms of *Agkistrodon* (copperheads and moccasins). Bull. Chic. Acad. Sci. 7(2):147-170.—Distribution, Morphology, Taxonomy. *Agkistrodon.*

700 **Gluesing, E. A.** 1983. Collared lizard predation: The effects of conspicuous morphology and movement. Copeia 1983(3):835-837.—Natural History, Techniques. *Crotaphytus.*

701 **Goin, C. J.** 1958. Notes on the maxillary dentition of some hylid frogs. Herpetologica 14(2):117-121.—Morphology. *Pseudacris.*

702 **Goin, C. J., and M. G. Netting.** 1940. A new gopher frog from the Gulf Coast, with comments upon the *Rana areolata* Group. Ann. Carnegie Mus. 28(28):137-168.—Distribution, Morphology, Taxonomy. *Rana.*

Goin, C. J., *see* 531

703 **Goldsmith, S. K.** 1981. Behavior and ecology of the rough green snake (*Opheodrys aestivus*). M.S. Thesis, Univ. Okla., Norman. 95 pp.—Behavior, Ecology. *Opheodrys.*

704 **Goldsmith, S. K.** 1984. Aspects of the natural history of the rough green snake, *Opheodrys aestivus* (Colubridae). Southwest. Nat. 29(4): 445-452.—Behavior, Ecology, Natural History. *Opheodrys.*

705 **Goldsmith, S. K.** 1986. Feeding behavior of an arboreal, insectivorous snake (*Opheodrys aestivus*) (Colubridae). Southwest. Nat. 31(2):246-249.—Behavior. *Opheodrys.*

Goode, J. M., *see* 478

706 **Gould, C. N.** 1929. Fossil bones and artifacts at Frederick. Proc. Okla. Acad. Sci. 9:90-92.—Paleontology. *Trionyx.*

707 **Graber, J. W.** 1957. A bioecological study of the black-capped vireo (*Vireo atricapillus*). Ph.D. Diss., Univ. Okla., Norman. 203 pp.—Black rat snakes were assumed predators of eggs. Natural History. *Elaphe.*

708 **Graber, J. W.** 1961. Distribution, habitat requirements, and life history of the black-capped vireo (*Vireo atricapillus*). Ecol. Monogr. 31(4):313-336.—Black rat snakes were assumed predators of eggs. Natural History. *Elaphe.*

709 **Graham, D. G.** 1965. Wildlife portraits: The crayfish frog. Okla. Wildl. 21(8):10-11.—Caption and photograph of crayfish frog. Popular. *Rana.*

710 **Graham, D. G.** 1966. Peeper. Outdoor Okla. 22(4):inside front cover.—Photograph and caption of spring peeper. Popular. *Hyla.*

711 **Graham, D. G.** 1967. Bullfrog. Outdoor Okla. 23(6):inside front cover.—Photograph with caption. Popular. *Rana.*

712 **Graham, D. G.** 1974. Outdoor Okla. 30(7):front cover.—Photograph of an Oklahoma eastern gray tree frog. Popular. *Hyla.*

713 **Graham, D. G.** 1980. Wildlife portrait: The copperhead. Outdoor Okla. 36(4):20.—Popular. *Agkistrodon.*

714 **Gratz, R. K.** 1973. *Elaphe guttata emoryi* (Great Plains rat snake). HISS News J. 1(3):99.—Distribution. *Elaphe.*

715 **Gratz, R. K.** 1973. *Thamnophis cyrtopsis cyrtopsis* (Western black-headed garter snake). HISS News J. 1(3):99.—Distribution. *Thamnophis.*

716 **Gratz, R. K.** 1976. Energetics for activity in the diamond-back water snake, *Natrix rhombifera.* Ph.D. Diss., Univ. Okla., Norman. 58 pp.—Physiology. *Nerodia.*

717 **Gratz, R. K.** 1978. Ventilation and gas exchange in the diamondback water snake, *Natrix rhombifera.* J. Comp. Physiol. 127(B):299-305.—Physiology. *Nerodia.*

718 **Gratz, R. K.** 1979. Ventilatory response of the diamondback water snake, *Natrix rhombifera,* to hypoxia, hypercapnia and increased oxygen demand. J. Comp. Physiol. 129(B):105-110.—Physiology. *Nerodia.*

719 **Gratz, R. K., and V. H. Hutchison.** 1977. Energetics for activity in the diamondback water snake. Physiol. Zool. 50(2):99-114.—Physiology. *Nerodia.*

720 **Greene, H. W., and G. V. Oliver, Jr.** 1965. Notes on the natural history of the western massasauga. Herpetologica 21(3):225-228.—Natural History. *Cnemidophorus, Sistrurus.*

Greene, H. W., *see* 639

721 **Greer, J. K., and R. L. Gilstrap.** 1970. Vertebrate remains in barn owl pellets. Bull. Okla. Ornithol. Soc. 3(4):25-29.—"Amphibians" in owl pellets. General Reference, Natural History.

722 **Gregory, J. U., F. E. Peabody, and L. I. Price.** 1956. Revision of the Gymnarthidae, American Permian microsaurs. Yale Univ. Peabody Mus. Nat. Hist. Bull. 10:1-77.—Included a fossil amphibian from Oklahoma. General Reference, Paleontology.

723 **Griffith, S. J.** 1966. Display variation and behavior associated with dominance in two subspecies of *Sceloporus undulatus*.-Iguanidae. Ph.D. Diss., Univ. Okla., Norman. 32 pp.—Behavior, Techniques. *Sceloporus.*

Grigsby, E. M., *see* 1402

724 **Grimpe, R. D.** 1984. Brown thrasher attacks lined snake. Bull. Okla. Herp. Soc. 9(3):27.—Natural History. *Tropidoclonion.*

725 **Grimpe, R. D.** 1984. Food induced combat in the western diamondback rattlesnake (*Crotalus atrox*). Bull. Okla. Herp. Soc. 9(1):1-3.—Behavior. *Crotalus.*

726 **Grimpe, R. D.** 1987. Maintenance, behavior and reproduction of the alligator snapping turtle, *Macrocleymys temmincki,* at the Tulsa Zoological Park. Bull. Okla. Herp. Soc. 12(1-4):1-6.—Husbandry. *Macroclemys.*

727 **Grimpe, R. D., and G. E. Benefield.** 1981. *Lampropeltis g. holbrooki* (speckled kingsnake). Reproduction. Herpetol. Rev. 12(3):88.—Natural History. *Lampropeltis.*

728 **Grimpe, R. D., J. Benefield, and J. Barkas.** 1979. Winter activity in two Oklahoma snakes. Bull. Okla. Herp. Soc. 4(2-3):45-46.—Natural History. *Lampropeltis, Nerodia.*

Grimpe, R. D., *see* 92, 1488

729 **Grobman, A. B.** 1944. The distribution of the salamanders of the genus *Plethodon* in eastern United States and Canada. Ann. N.Y. Acad. Sci. 45(7):261-316.—Distribution. *Plethodon.*

730 **Grobman, A. B.** 1950. The distribution of the races of *Desmognathus fuscus* in the southern states. Nat. Hist. Misc. (Chic.) (20):1-8.—Distribution. *Desmognathus.*

731 **Grobman, A. B.** 1984. Scutellation variation in *Opheodrys aestivus.* Bull. Fla. State Mus. Biol. Sci. 29(3):153-170.—Distribution, Taxonomy. *Opheodrys.*

Grubitz, III, G., *see* 480

732 **Guberlet, J. E.** 1920. A new bladder fluke from the frog. Trans. Am. Microsc. Soc. 39(2):142-148.—Parasitology. *Rana.*

Gumbreck, L. G., *see* 5

Guttman, S. I., *see* 1050, 1346

733 **Haglund, T. R.** 1977. New occurrences and paleoecology of *Peronedon primus* Olson (Nectridea). J. Paleontol. 51(5):982-985.—General Reference, Paleontology.

734 **Hahn, D. E.** 1979. *Leptotyphlops.* Cat. Am. Amphib. Reptiles 230.1-230.4.—Range map and literature review. Bibliographies, Distribution. *Leptotyphlops.*

735 **Hahn, D. E.** 1979. *Leptotyphlops dulcis.* Cat. Am. Amphib. Reptiles 231.1-231.2.—Range map and literature review. Bibliographies, Distribution. *Leptotyphlops.*

736 **Hall, R. J.** 1976. *Eumeces obsoletus.* Cat. Am. Amphib. Reptiles 186.1-186.3.—Range map and literature review. Bibliographies, Distribution. *Eumeces.*

Hall, S. A., *see* 756

737 **Hallowell, E.** 1852. Descriptions of new species of reptiles inhabiting North America. Proc. Acad. Nat. Sci. Phila. 6:177-182.—General Reference, Historical.

738 **Hallowell, E.** 1854. Reptiles. Pages 106-147 *in* L. Sitgreaves. Report of an expedition down the Zuni and Colorado Rivers in 1851. 33rd. U.S. Cong. 1st. Ses. Sen. Exec. Doc. Wash. 198 pp.—General Reference, Historical.

739 **Hallowell, E.** 1857. Notice of a collection of reptiles from Kansas and Nebraska, presented to the Academy of Natural Sciences, by Dr. Hammond. Proc. Acad. Nat. Sci. Phila. 8:238-253.—General Reference, Historical.

740 **Hannum, C. A.** 1925. A new species of cestode, *Ophiotaenia magna* n. sp. from the frog. Trans. Am. Microsc. Soc. 44(3):148-152.—Parasitology. *Rana.*

Hardy, L. M., *see* 1227

741 **Harms, J. H.** 1938. The embryological development of the urogenital system in *Scaphiopus hammondi* Baird. M.S. Thesis, Univ. Okla., Norman. 53 pp.—Development. *Scaphiopus.*

742 **Harney, L.** 1953. Notes on the food habits of three common lizards of southern Oklahoma. Proc. Okla. Acad. Sci. 34:85-86.—Ecology, Natural History. *Sceloporus, Scincella, Cnemidophorus.*

743 **Harper, F.** 1947. A new cricket frog (*Acris*) from the middle western states. Proc. Biol. Soc. Wash. (60):39-40.—Distribution, Taxonomy. *Acris.*

Harrell, B. E., *see* 993

Harvey, B. C., *see* 1178

744 **Harwood, P. D.** 1931. Some parasites of Oklahoma turtles. J. Parasitol. 18(2):98-101.—Parasitology. *Graptemys, Trachemys.*

745 **Hay, O. P., and H. J. Cook.** 1930. Fossil vertebrates collected near, or in association with, human artifacts at localities near Colorado, Texas; Frederick, Oklahoma; and Folsom, New Mexico. Proc. Colo. Mus. Nat. Hist. 9(2):1-55.—Distribution, Paleontology. *Trionyx.*

Hayes, S., *see* 370

746 **Hazard, E. S.** 1977. Ontogenetic changes in erythrocytic organic phosphates in the bullfrog *Rana catesbeiana.* M.S. Thesis, Univ. Okla., Norman. 23 pp.—Development, Physiology. *Rana.*

747 **Hazard, E. S.** 1981. Erythrocytic phosphates in amphibians. Ph.D. Diss., Univ. Okla., Norman. 99 pp.—Physiology. *Bufo, Pseudacris, Scaphiopus, Rana.*

748 **Hazard, E. S., and V. H. Hutchison.** 1978. Ontogenetic changes in erythrocytic organic phosphates in the bullfrog, *Rana catesbeiana.* J. Exp. Zool. 206(1):109-118.—Development, Physiology. *Rana.*

749 **Hazard, E. S., and V. H. Hutchison.** 1982. Distribution of acid soluble phosphates in the erythrocytes of selected species of amphibians. Comp. Biochem. Physiol. 73(A):111-125.—Physiology. *Bufo, Hyla, Pseudacris, Scaphiopus.*

750 **Heaton, M. J.** 1979. Cranial anatomy of primitive captorhinid reptiles from the Late Pennsyvanian and the Early Permian of Oklahoma and Texas. Okla. Geol. Surv. Bull. 127:1-84.—General Reference, Paleontology.

751 **Heaton, M. J., and R. R. Reisz.** 1980. A skeletal reconstruction of the early Permian captorhinid reptile *Eucaptorhinus laticeps* (Williston). J. Paleontol. 54(1):136-143.—General Reference, Paleontology.

752 **Hecht, M. K.** 1958. A synopsis of the mudpuppies of eastern North America. Proc. Staten Island Inst. Arts Sci. 21(1):1-37.—Bibliographies, Distribution, Taxonomy. *Necturus.*

753 **Hecht, M. K., and B. L. Matalas.** 1946. A review of middle North America toads of the genus *Microhyla.* Am. Mus. Novit. (1315):1-21.—Distribution, Taxonomy. *Gastrophryne.*

754 **Heck, D. D., Jr.** 1976. Alligator snapping turtle from Lake Eufaula. Bull. Okla. Herp. Soc. 1(3):34.—Distribution. *Macroclemys.*

Hefley, H. M., *see* 1422

Heinze, A. A., *see* 606

755 **Heinzig, F. F.** 1950. A cytological analysis of comparative growth rates in *Scaphiopus hurteri* and *Rana pipiens.* M.S. Thesis, Univ. Okla., Norman. 48 pp.—Development, Morphology. *Scaphiopus, Rana.*

Hellbaum, A. A., *see* 1085

756 **Henry, D. O., B. H. Butler, and S. A. Hall.** 1979. The late prehistoric human ecology of Birch Creek Valley, northern Oklahoma. Plains Anthrop. 24(85):207-238.—General Reference, Paleontology.

757 **Hensley, M.** 1959. Albinism in North American amphibians and reptiles. Publ. Mus. Mich. State Univ. Biol. Surv. 1(4):135-159.—Morphology. *Lampropeltis.*

Hepworth, L., *see* 106

758 **Hesse, C. J.** 1936. Lower Pliocene vertebrate fossils from the Ogallala Formation (Lavern Zone) of Beaver County, Oklahoma. Carnegie Inst. Wash. Publ. (476):1-72.—Tortoise and snake fossils included. General Reference, Paleontology.

759 **Hibbard, C. W.** 1954. A new Pliocene vertebrate fauna from Oklahoma. Paps. Mich. Acad. Sci. Arts. Lett. (39):339-359.—General Reference, Paleontology.

Hibbard, C. W., *see* 1235

760 **Hicks, L.** 1971. Wildlife portrait: Black rat snake. Outdoor Okla. 27(7):20.—Popular. *Elaphe.*

761 **Hicks, L.** 1971. Wildlife portraits: The copperhead. Outdoor Okla. 27(1):20.—Photograph included. Popular. *Agkistrodon.*

762 **Hicks, L.** 1971. Wildlife portrait: Tiger salamander. Outdoor Okla. 27(6):20.—Photograph included. Popular. *Ambystoma.*

763 **Hicks, L.** 1972. Wildlife portrait: Rattlesnake. Outdoor Okla. 28(5): 20.—An article about timber rattlesnakes in Oklahoma. Popular. *Crotalus.*

764 **Hicks, L.** 1972. Wildlife portrait: The toad. Outdoor Okla. 28(3): 20.—An article about American toads in Oklahoma. Popular. *Bufo.*

765 **Hicks, L.** 1973. Wildlife portrait: The box turtle. Outdoor Okla. 29(4):20.—Incorrectly printed as volume 28. Popular. *Terrapene.*

766 **Hicks, L.** 1973. Wildlife portrait: Leopard frog. Outdoor Okla. 29(5):20.—Popular. *Rana.*

767 **Hicks, L.** 1974. Conservation notebook: Oklahoma wildlife symbols. Outdoor Okla. 30(2):centerfold.—Popular. *Crotaphytus.*

768 **Hicks, L.** 1975. Wildlife portrait: The copperhead. Outdoor Okla. 31(6):20.—Photograph included inside front cover. Popular. *Agkistrodon.*

769 **Hicks, L.** 1976. Wildlife portrait: The black rat snake. Outdoor Okla. 32(5):20.—Photograph included inside front cover. Popular. *Elaphe.*

770 **Hicks, L.** 1976. Wildlife portrait: Tiger salamander. Outdoor Okla. 32(2):20.—Photograph included inside front cover. Popular. *Ambystoma.*

Higginbotham, J. W., *see* 798

771 **Highton, R.** 1960. Heritability of geographic variation in trunk segmentation in the red-backed salamander, *Plethodon cinereus.* Evolution 14(3):351-360.—Distribution, Morphology. *Plethodon.*

772 **Highton, R.** 1962. Revision of North American salamanders of the genus *Plethodon.* Bull. Fla. State Mus. Biol. Sci. 6(3):235-267.—Bibliographies, Distribution, Ecology, Taxonomy. *Plethodon.*

773 **Highton, R.** 1986. *Plethodon serratus.* Cat. Am. Amphib. Reptiles 394.1-394.2.—Range map and literature review. Bibliographies, Distribution. *Plethodon.*

774 **Highton, R., and A. Larson.** 1979. The genetic relationships of salamanders of the genus *Plethodon.* Syst. Zool. 28(4):579-599.—*Plethodon.*

775 **Highton, R., and T. P. Webster.** 1976. Geographic protein variation and divergence in populations of the salamander *Plethodon cinereus.*

Evolution 30(1):33-35.—Distribution, Evolution, Genetics, Taxonomy. *Plethodon.*

Highton, R., *see* 588, 949

Hill, L. G., *see* 816

776 **Hill, W. C.** 1941. *Physaloptera terrapenis,* a new trematode from a tortoise. Trans. Am. Microsc. Soc. 60(1):59-64.—Parasitology. *Terrapene.*

777 **Hillis, D. M.** 1976. A hybrid *Eumeces* from Oklahoma. Bull. Md. Herpetol. Soc. 11(4):177-178.—Taxonomy. *Eumeces.*

778 **Hillis, D. M.** 1981. Premating isolating mechanisms among three species of the *Rana pipiens* complex in Texas and southern Oklahoma. Copeia 1981(2):312-319.—Behavior, Distribution, Evolution. *Rana.*

779 **Hillis, D. M., J. S. Frost, and D. A. Wright.** 1983. Phylogeny and biogeography of the *Rana pipiens* complex: A biochemical evaluation. Syst. Zool. 32(2):132-143.—Distribution, Taxonomy. *Rana.*

780 **Hinde, H. P.** 1940. Cytoplasmic inclusions in the embryos of *Bufo woodhousei woodhousei* Girard. M.S. Thesis, Univ. Okla., Norman. 61 pp.—Development. *Bufo.*

Hoestenbach, R. D., Jr., *see* 1444

781 **Holle, D. G.** 1973. Diet and general availability of prey of the coyote (*Canis latrans*) at the Wichita Mountains National Wildlife Refuge, Oklahoma. M.S. Thesis, Okla. State Univ., Stillwater. 59 pp.—Collared lizards, snakes and other reptiles were found in scats. Natural History. *Ophisaurus, Crotaphytus.*

782 **Hollis, P. D.** 1979. The histochemical demonstration of alkaline phosphatase activity in *Porocephalus crotali* Humbolt, 1811 (Pentastomida). J. Parasitol. 65(3):461-463.—Parasitology. *Crotalus.*

783 **Hollis, P. D.** 1979. The neuroanatomy of *Porocephalus crotali* Humboldt, 1811 (Pentastomida). Trans. Am. Microsc. Soc. 98(1):114-122.—Parasitology. *Crotalus.*

784 **Holman, J. A.** 1971. *Ophisaurus.* Cat. Am. Amphib. Reptiles 110.1-110.3.—Range map and literature review. Bibliographies, Distribution. *Ophisaurus.*

785 **Holman, J. A.** 1971. *Ophisaurus attenuatus.* Cat. Am. Amphib. Reptiles 111.1-111.3.—Range map and literature review. Bibliographies, Distribution. *Ophisaurus.*

786 **Holman, J. A.** 1971. *Ophisaurus ventralis.* Cat. Am. Amphib. Reptiles 115.1-115.2.—This species may occur in Oklahoma, but is not confirmed. Range map and literature review. Distribution. *Ophisaurus.*

787 **Holman, J. A.** 1973. A new Pliocene snake, genus *Elaphe,* from Oklahoma. Copeia 1973(3):574-580.—Paleontology. *Elaphe.*

788 **Holman, J. A.** 1986. Snakes of the Berends Local Fauna (Pleistocene: Early Illinoian) of Oklahoma. Copeia 1986(3):811-812.—Paleontology. *Trachemys, Kinosternon, Sternotherus, Elaphe, Coluber, Heterodon, Lampropeltis, Masticophis, Nerodia, Pituophis, Thamnophis.*

789 **Holmes, R.** 1977. The osteology and musculature of the pectoral limb of small captorhinids. J. Morphol. 152(1):101-140.—The fossil reptile was from Oklahoma. General Reference, Morphology, Paleontology.

790 **Holzapfel, R. A.** 1935. The cyclic character of hibernation in frogs. Ph.D. Diss., Univ. Okla., Norman. 204 pp.—Physiology. *Acris, Pseudacris, Gastrophryne, Rana.*

791 **Holzapfel, R. A.** 1937. Cyclic character of hibernation of frogs. Quart. Rev. Biol. 12(1):65-84—Physiology. *Acris, Pseudacris, Gastrophryne, Rana.*

Hoover, J. J., *see* 95

792 **Houseal, T. W., J. W. Bickham, and M. D. Springer.** 1982. Geographic variation in the yellow mud turtle, *Kinosternon flavescens.* Copeia 1982(3):567-580.—Distribution, Morphology, Taxonomy. *Kinosternon.*

793 **Howell, P.** 1978. Swamplands. Outdoor Okla. 34(5):12-17.—Popular. *Trachemys, Eumeces.*

Hoyle, W. L., *see* 423

794 **Hubbard, W. E.** 1938. *Ophiopneumicola colubri* N. M., n. sp. a lung mite from a snake. Trans. Am. Microsc. Soc. 57(4):400-406.—Parasitology. *Masticophis.*

795 **Hubbard, W. E.** 1939. *Eutonyssus ewingi* n. sp., an ophidian lung mite. Am. Midl. Nat. 21(3):657-662.—Parasitology. *Agkistrodon.*

796 **Hubbard, W. E.** 1940. Observations on *Entonyssus ewingi* Hubbard 1939. Trans. Am. Microsc. Soc. 59(2):243-247.—Parasitology. *Crotalus.*

Hudson, W. F., *see* 371, 372

797 **Huffman, G. C., T. A. Cathey, and J. E. Humphrey.** 1963. A guide to the state parks and scenic areas in the Oklahoma Ozarks. Okla. Geol. Surv. Guidebook 12. 95 pp.—Distribution. *Bufo, Acris, Hyla, Pseudacris, Gastrophryne, Rana, Ambystoma, Eurycea, Necturus, Chelydra, Chrysemys, Deirochelys, Graptemys, Pseudemys, Trachemys, Terrapene, Kinosternon, Sternotherus, Trionyx, Ophisaurus, Crotaphytus, Phrynosoma, Sceloporus, Eumeces, Scincella, Cnemidophorus, Carphophis, Cemophora, Coluber, Diadophis, Elaphe, Heterodon, Lampropeltis, Masticophis, Nerodia, Opheodrys, Pituophis, Storeria, Thamnophis, Virginia, Agkistrodon, Crotalus, Sistrurus.*

Hughes, R. C., *see* 798, 1073, 1074

798 **Hughes, R. C., J. W. Higginbotham, and J. W. Clary.** 1942. The trematodes of reptiles, Part I. Systematic section. Am. Midl. Nat. 27(1):109-134.—Parasitology. *Trachemys, Heterodon.*

799 **Hughes, R. C., and G. A. Moore.** 1943. *Acanthocephalus van-cleavei,* a new echinorhynchid worm, from a salamander. Am. Midl. Nat. 29(3):724-729.—Parasitology. *Eurycea.*

800 **Hughes, W.** 1947. Collared lizard. Okla. Game Fish News 3(6):14.—Photograph included. Popular. *Crotaphytus.*

801 **Hughes, W.** 1947. Friend to man-the toad. Okla. Game Fish News 3(9):14.—Popular. *Bufo.*

802 **Hughes, W.** 1947. Rattlesnake roundup. Okla. Game Fish News 3(5):4-5, 16.—Popular. *Crotalus.*

803 **Hughes, W.** 1948. Albinistic animals. Okla. Game Fish News 4(2): 6.—Popular. *Elaphe.*

804 **Hughes, W.** 1950. Exit-rattlers! Okla. Game Fish News 6(5):12-13.—Popular. *Crotalus.*

805 **Hughes, W.** 1950. Reptile review. Okla. Game Fish News 6(3):12-13.—Photographs. Popular. *Terrapene, Ophisaurus, Crotaphytus, Agkistrodon, Crotalus.*

806 **Hughes, W.** 1954. Hallowe'en spookface on the cover. Okla. Game Fish News 10(10):front cover, 10.—Popular. *Chelydra.*

807 **Hughes, W.** 1955. Animal faces. Okla. Game Fish News 11(3):12.—Popular. *Bufo, Terrapene, Pituophis.*

808 **Hughes, W.** 1955. Interesting amphibian. Okla. Game Fish News 11(7-8):15.—Natural History, Popular. *Ambystoma.*

809 **Hughes, W.** 1955. Signs of spring. Okla. Game Fish News 11(3): 16.—Popular. *Rana.*

810 **Huheey, J. E.** 1959. Distribution and variation in the glossy snake, *Natrix rigida* (Say). Copeia 1959(4):303-311.—Distribution, Taxonomy. *Regina.*

Humphrey, J. E., *see* 797

811 **Hunkapiller, T.** 1976. Turtle-eating black rat snake in eastern Oklahoma. Bull. Okla. Herp. Soc. 1(4):61.—Natural History. *Trachemys, Elaphe.*

Hunkapiller, T., *see* 153

812 **Hurter, J., and J. K. Strecker.** 1909. Amphibians and reptiles of Arkansas. Trans. Acad. Sci. St. Louis 18(2):11-27.—Oklahoma specimens listed. Distribution, Historical. *Eurycea, Ophisaurus, Phrynosoma, Sceloporus, Eumeces, Carphophis, Elaphe, Heterodon, Lampropeltis, Masticophis, Opheodrys, Virginia, Agkistrodon.*

Hussakof, L., *see* 495

813 **Hutchison, V. H.** 1956. Notes on the plethodontid salamanders, *Eurycea lucifuga* (Rafinesque) and *E. longicauda longicauda* (Green). Occas. Pap. Nat. Speleol. Soc. 3:1-24.—Distribution, Morphology, Natural History, Taxonomy. *Eurycea.*

814 **Hutchison, V. H.** 1958. The distribution and ecology of the cave salamander, *Eurycea lucifuga.* Ecol. Monogr. 28(1):1-20.—Bibliographies, Distribution, Ecology. *Eurycea.*

815 **Hutchison, V. H.** 1966. *Eurycea lucifuga.* Cat. Am. Amphib. Reptiles 24.1-24.2.—Range map and literature review. Bibliographies, Distribution. *Eurycea.*

816 **Hutchison, V. H., and L. G. Hill.** 1978. Thermal selection of bullfrog (*Rana catesbeiana*) tadpoles at different stages of development and acclimation temperatures. J. Thermal Biol. 3(1):57-60.—Development, Physiology. *Rana.*

817 **Hutchison, V. H., and K. Miller.** 1979. Anaerobic capacity in amphibians. Comp. Biochem. Physiol. 63(A):213-216.—Physiology. *Hyla, Rana, Scaphiopus.*

818 **Hutchison, V. H., A. Vinegar, and R. J. Kosh.** 1966. Critical thermal maxima in turtles. Herpetologica 22(1):32-41.—Physiology. *Trachemys.*

Hutchison, V. H., *see* 541, 618, 719, 748, 749, 1180, 1181

819 **Ireland, P. H.** 1973. Marking larval salamanders with fluorescent pigments. Southwest. Nat. 18(2):241-245.—Techniques. *Ambystoma, Eurycea.*

820 **Ireland, P. H.** 1979. *Eurycea longicauda.* Cat. Am. Amphib. Reptiles 221.1-221.4.—Range map and literature review. Bibliographies, Distribution. *Eurycea.*

821 **Ireland, P. H., and R. Altig.** 1983. Key to the gilled salamander larvae and larviform adults of Arkansas, Kansas, Missouri, and Oklahoma. Southwest. Nat. 28(3):271-274.—Taxonomy. All Caudata.

Ireland, P. H., *see* 8

822 **Irving, W.** 1956. A tour of the prairies. Univ. Okla. Press, Norman. 214 pp.—Historical. *Crotalus.*

823 **Irwin, K. J.** 1979. Black Mesa meeting a huge success. Kans. Herpetol. Soc. Newsl. (31):1-2.—Distribution. *Bufo, Gastrophryne, Scaphiopus, Rana, Terrapene, Kinosternon, Trionyx, Crotaphytus, Holbrookia, Phrynosoma, Sceloporus, Eumeces, Cnemidophorus, Diadophis, Elaphe, Heterodon, Hypsiglena, Masticophis, Pituophis, Sonora, Tantilla, Thamnophis, Tropidoclonion, Leptotyphlops, Crotalus.*

824 **Iverson, J. B.** 1977. *Kinosternon subrubrum.* Cat. Am. Amphib. Reptiles 193.1-193.4.—Range map and literature review. Distribution. *Kinosternon.*

825 **Iverson, J. B.** 1979. *Sternotherus carinatus.* Cat. Am. Amphib. Reptiles 226.1-226.2.—Range map and literature review. Bibliographies, Distribution. *Sternotherus.*

826 **Iverson, J. B.** 1979. A taxonomic reappraisal of the yellow mud turtle, *Kinosternon flavescens* (Testudines: Kinosternidae). Copeia 1979(2):212-225.—Taxonomy. *Kinosternon.*

827 **Iverson, J. B.** 1982. Biomass in turtle populations: A neglected subject. Oecologia 55(1):69-76.—Ecology. *Kinosternon, Sternotherus.*

828 **Iverson, J. B.** 1985. Checklist of the turtles of the world with English common names. Soc. Study Amphib. Reptiles Herpetol. Circ. (14):1-14.—Taxonomy. All Testudines.

829 **Iverson, J. B., and S. A. Iverson.** 1980. A bibliography to the mud and musk turtle family Kinosternidae. Smithson. Herpetol. Infor. Serv. (48):1-72.—Bibliographies. *Kinosternon, Sternotherus.*

Iverson, S. A., *see* 829

830 **Jackson, J. A.** 1976. Relative climbing tendencies of gray (*Elaphe obsoleta spiloides*) and black rat snakes (*E. o. obsoleta*). Herpetologica 32(4):359-361.—Some of the black rat snakes were from southeast Oklahoma. Behavior. *Elaphe.*

831 **Jackson, M. K.** 1977. Histology and distribution of cutaneous touch corpuscles in some leptotyphlopid and colubrid snakes (Reptilia, Serpentes). J. Herpetol. 11(1):7-15.—Specimens were from the Stovall museum. Morphology. *Leptotyphlops.*

832 **Jackson, M. K., and H. W. Reno.** 1975. Comparative skin structure of some fossorial and subfossorial leptotyphlopid and colubrid snakes. Herpetologica 31(3):350-359.—Morphology. *Leptotyphlops.*

833 **James, E.** 1823. Account of an expedition from Pittsburgh to the Rocky Mountains, performed in the years 1819 and 1820. H. C. Carey and Lea, Philadelphia. 442 pp.—General Reference, Historical, Taxonomy. *Crotaphytus.*

834 **Jeffs, R. E.** 1926. Albert Heald Van Vleet, a pioneer Oklahoma scientist. Proc. Okla. Acad. Sci. 6:379-382.—An obituary and historical account of Van Vleet's contributions to Oklahoma herpetology. General Reference, Historical.

835 **Jenni, B.** 1954. Two tails odd but not a freak. Okla. Game Fish News 10(9):11.—Popular. *Cnemidophorus.*

836 **Jenni, B.** 1954. Watch those dry leaves! Okla. Game Fish News 10(6):20.—Popular. *Agkistrodon.*

837 **Jenni, B.** 1955. Wake up! Its 58 degrees fahrenheit. Okla. Game Fish News 11(3):10.—Popular. *Masticophis, Agkistrodon, Crotalus.*

838 **Jenni, B.** 1958. Look out rattler! Okla. Wildl. 14(4):9.—Story on rattlesnake hunts. Popular. *Crotalus.*

839 **Jenni, B.** 1964. Snakes like it warm. Okla. Wildl. 20(12):16-17.—Popular. *Agkistrodon, Crotalus.*

840 **Jenni, B.** 1965. Naturalist's notebook. Okla. Wildl. 21(3):20-21.—Discussion on water moccasins. Popular. *Agkistrodon.*

841 **Jenni, B.** 1965. Naturalist's notebook. Okla. Wildl. 21(4):20-21.—Discussion on rattlesnakes. *Crotalus, Sistrurus.*

842 **Jenni, B.** 1965. Naturalist's notebook. Okla. Wildl. 21(2):14-15.—Discussion on copperheads. Popular. *Agkistrodon.*

843 **Jenni, B.** 1966. Snake dance. Outdoor Okla. 22(5):6-7.—Describes combat dance of diamondback rattlesnakes. Popular. *Crotalus.*

844 **Jenssen, T. A., and W. B. Preston.** 1968. Behavioral responses of the male green frog, *Rana clamitans,* to its recorded call. Herpetologica 24(2):181-182.—Behavior. *Rana.*

845 **Johnson, C.** 1963. Additional evidence of sterility between call-types in the *Hyla versicolor* complex. Copeia 1963(1):139-143.—Behavior, Natural History, Taxonomy. *Hyla.*

846 **Johnson, C.** 1966. Species recognition in the *Hyla versicolor* complex. Tex. J. Sci. 18(4):361-364.—Behavior. *Hyla.*

Johnson, C., *see* 968

Johnson, F. L., *see* 1272

847 **Johnson, V. O.** 1939. A supplementary note on the larvae of *Bufo woodhousei woodhousei.* Herpetologica 1(6):160-164.—Morphology. *Bufo.*

848 **Johnson, V. O.** 1940. Organization and differentiation in the early embryos of *Bufo woodhousei* Girard. M.S. Thesis, Univ. Okla., Norman. 159 pp.—Development. *Bufo.*

849 **Jones, J. P.** 1926. The proper name for *Sceloporus consobrinus* Baird and Girard. Occas. Pap. Mus. Zool. Univ. Mich. (172):1-3.—Historical, Taxonomy. *Sceloporus.*

850 **Jones, R. W., and G. E. Derrick.** 1938. A new record of *Bufo punctatus* in Murray County, Oklahoma. Proc. Okla. Acad. Sci. 19: 39.—Distribution, Natural History. *Bufo.*

851 **Jordan, J. A.** 1979. Box turtles and stomp dancing. Stovall Mus. Newsl. 5(1):1-2.—Lore, Historical, Popular. *Terrapene.*

Jordan, J. E., *see* 951

Karges, J. P., *see* 1033

852 **Kassing, E. F.** 1961. A life history study of the Great Plains ground snake, *Sonora episcopa episcopa* (Kennicott). Tex. J. Sci. 13(2):185-203.—Behavior, Ecology, Natural History. *Sonora.*

853 **Keeton, D., and C. C. Carpenter.** 1955. A study of the variations of dorsal markings of *Bufo w. woodhousei.* Proc. Okla. Acad. Sci. 36:81-83.—Morphology. *Bufo.*

854 **Kemmerer, M. R., and G. R. Cline.** 1985. *Scincella lateralis* (ground skink). Predation. Herpetol. Rev. 16(1):27-28.—Natural History. *Scincella.*

855 **Kern, G. W.** 1956. A shrew eaten by a king snake, *Lampropeltis calligaster* (Harlan). Herpetologica 12(2):135.—A prairie kingsnake was observed eating a "*Blarina*" near Ardmore. Natural History. *Lampropeltis.*

856 **Kezer, J.** 1952. Thyroxin induced metamorphosis of the neotenic salamanders *Eurycea tynerensis* and *Eurycea neotenes.* Copeia 1952(4): 234-237.—Specimens were provided by G. A. Moore. Physiology. *Eurycea.*

857 **Kiester, A. R.** 1971. Species density of North American amphibians and reptiles. Syst. Zool. 20(2):127-137.—Distribution, Ecology. General Reference.

858 **Kiester, L.** 1968. Trotline turtles taste terrific. Outdoor Okla. 24(9): 13.—Popular. *Chelydra, Trionyx.*

859 **Kilgore, D. L.** 1969. An ecological study of the swift fox (*Vulpes velox*) in the Oklahoma Panhandle. Am. Midl. Nat. 81(2):512-534.—Great Plains toad found in one fox stomach. Natural History. *Bufo.*

860 **Kimsey, A.** 1952. The snake with a noisy tail. Okla. Game Fish News 8(6):8-10.—Popular. *Masticophis, Crotalus.*

861 **Kimsey, A.** 1955. Game and fish department aids town under water. Okla. Game Fish News 11(7-8):8-9.—Photograph included of bullsnake. Popular. *Pituophis.*

862 **Kimsey, A.** 1956. Okla. Game Fish News 12(6):15.—Photograph of albino diamondback rattlesnake. Morphology, Popular. *Crotalus.*

863 **Kimsey, A.** 1956. Cottonmouths crawl at night. Okla. Game Fish News 12(9):4-5.—Popular. *Agkistrodon.*

King, F. W., *see* 101

864 **King, O. M.** 1960. Observations on Oklahoma toads. Southwest. Nat. 5(2):103.—Issue contents incorrectly lists A. N. Bragg as author. Behavior. *Bufo, Scaphiopus.*

King, O. M., *see* 373

Kingsinger, R., Jr., *see* 375

865 **Kirn, A. J., W. L. Burger, Jr., and H. M. Smith.** 1949. The subspecies of Tantilla gracilis. Am. Midl. Nat. 42(1):238-251.—Distribution, Taxonomy. *Tantilla.*

866 **Kitts, D. B.** 1959. Cenozoic geology of northern Roger Mills County Oklahoma. Okla. Geol. Surv. Circ. (48):1-26.—*Testudo* also noted. Paleontology. *Coluber.*

867 **Kitts, D. B., and C. C. Black.** 1959. A Pliocene vertebrate local fauna from Roger Mills County, Oklahoma. Okla. Geol. Surv. Circ. (48): 27-47.—*Testudo* also noted. Paleontology. *Coluber.*

868 **Klau, H. H.** 1948. A biometric investigation of certain morphologic characters in *Heterodon contortrix contortrix* Linnaeus and *Coluber flagellum testaceus* Say: Length, ventral scale number, pyloric position and reduction position (first bilateral) of dorsal scale rows. M.S. Thesis, Univ. Okla., Norman. 272 pp.—Morphology. *Heterodon, Masticophis.*

869 **Klau, H. H., and P. R. David.** 1952. The bimodality of length distribution in *Heterodon p. platyrhinos* L. and its relation to the season in which the specimens were collected. Am. Midl. Nat. 47(2):364-371.—Morphology, Natural History. *Heterodon.*

870 **Klauber, L. M.** 1930. Differential characteristics of southwestern rattlesnakes allied to *Crotalus atrox.* Bull. Zool. Soc. San Diego (6):1-72.—Distribution, Morphology. *Crotalus.*

871 **Klauber, L. M.** 1936. A key to the rattlesnakes with summary of characteristics. Trans. San Diego Soc. Nat. Hist. 8(20):185-276.—Distribution, Taxonomy. *Crotalus, Sistrurus.*

872 **Klauber, L. M.** 1940. The worm snakes of the genus *Leptotyphlops* in the United States and northern Mexico. Trans. San Diego Soc. Nat. Hist. 9(18):87-160.—Distribution, Morphology, Taxonomy. *Leptotyphlops.*

873 **Klauber, L. M.** 1941. The long-nosed snakes of the genus *Rhinocheilus.* Trans. San Diego Soc. Nat. Hist. 9(29):289-332.—Distribution, Taxonomy. *Rhinocheilus.*

874 **Klauber, L. M.** 1946. The glossy snake, *Arizona,* with descriptions of new subspecies. Trans. San Diego Soc. Nat. Hist. 10(17):311-398.—Distribution, Taxonomy. *Arizona.*

875 **Klauber, L. M.** 1947. Classification and ranges of the gopher snakes of the genus *Pituophis* in the western United States. Bull. Zool. Soc.

San Diego (22):1-83.—Distribution, Morphology, Taxonomy. *Pituophis.*

876 **Klauber, L. M.** 1956. Rattlesnakes: Their habits, life histories, and influence of mankind. 2 volumes. Univ. California Press, Berkeley, California, 1476 pp.—Bibliographies, Distribution, Lore, Natural History, Taxonomy. *Crotalus, Sistrurus.*

Klicka, J., *see* 990

877 **Kluge, A. G.** 1983. Type-specimens of amphibians in the University of Michigan Museum of Zoology. Misc. Publ. Mus. Zool. Univ. Mich. (166):1-68.—Distribution, Taxonomy. *Eurycea, Typhlotriton.*

Knight, J. L., *see* 522

878 **Kolb, D. W.** 1946. Snake story. Okla. Game Fish News 2(10):22.—Natural History, Popular. *Crotalus.*

Kosh, R. J., *see* 818

879 **Krojnovich, R.** 1971. Early Oklahoma "wildlife." Outdoor Okla. 27(11):2-5.—General Reference, Paleontology, Popular.

880 **Kroll, J. C.** 1973. Taste buds in the oral epithelium of the blind snake, *Leptotyphlops dulcis* (Reptilia: Leptotyphlopidae). Southwest. Nat. 17(4):365-370.—Specimens were collected in Oklahoma. Morphology. *Leptotyphlops.*

881 **Kroll, J. C., and H. W. Reno.** 1971. A re-examination of the cloacal sacs and gland of the blind snake, *Leptotyphlops dulcis* (Reptilia: Leptotyphlopidae). J. Morphol. 133(3):273-280.—Morphology. *Leptotyphlops.*

Kroll, J. C., *see* 679, 1469

882 **Krupa, J. J.** 1985. *Sceloporus undulatus erythrocheilus* (red-lipped lizard). Injuries. Herpetol. Rev. 16(2):54.—Frequency of cactus spine injuries. Natural History. *Sceloporus.*

883 **Krupa, J. J.** 1986. Anuran breeding dates in central Oklahoma. Bull. Okla. Herp. Soc. 11(1-4):10-13.—Ecology. *Bufo, Acris, Hyla, Pseudacris, Gastrophryne, Scaphiopus.*

884 **Krupa, J. J.** 1986. Distribution in Oklahoma of the bird-voiced treefrog (*Hyla avivoca*). Proc. Okla. Acad. Sci. 66:37-38.—Distribution. *Hyla.*

885 **Krupa, J. J.** 1986. Multiple egg clutch production in the Great Plains toad. Prairie Nat. 18(3):151-152.—Ecology, Natural History. *Bufo.*

886 **Krupa, J. J.** 1986. Perch preference by the fence lizard (*Sceloporus undulatus*) under experimental conditions. Bull. Okla. Herp. Soc. 11(1-4):6-9.—Behavior, Ecology. *Sceloporus.*

887 **Krupa, J. J.** 1987. Mate choice and mate location tactics in the Great Plains toad (*Bufo cognatus*). Ph.D. Diss., Univ. Okla., Norman, 90 pp.—Behavior. *Bufo, Scaphiopus.*

888 **Krupa, J. J., S. M. Secor, and G. Sievert.** 1985. Observations of the bird-voiced treefrog in Oklahoma: A resighting of an isolated population after 23 years. Bull. Okla. Herp. Soc. 10(1-2):8-10.—Distribution, Natural History. *Hyla.*

889 **Kuntz, R. E.** 1940. *Hypsiglena ochrorhyncha* in Oklahoma. Copeia 1940(2):136.—Distribution. *Hypsiglena.*

890 **Kuntz, R. E.** 1940. The metazoan parasites of some Oklahoma Anura. Proc. Okla. Acad. Sci. 21:33-34.—Parasitology. *Bufo, Acris, Pseudacris, Gastrophryne, Scaphiopus, Rana.*

891 **Kuntz, R. E.** 1940. A study of anuran parasites of Comanche County, Oklahoma. M.S. Thesis, Univ. Okla., Norman. 93 pp.—Ecology, Parasitology. All Anura.

892 **Kuntz, R. E., and J. T. Self.** 1943. An ecological study of the metazoan parasites of the Salientia of Comanche County, Oklahoma. Proc. Okla. Acad. Sci. 24:35-38.—Parasitology. *Bufo, Acris, Pseudacris, Gastrophryne, Scaphiopus, Rana.*

Kuntz, R. E., *see* 374, 1238, 1295

893 **Landreth, H. F.** 1972. Physiological responses of *Elaphe obsoleta* and *Pituophis melanoleucus* to lowered ambient temperatures. Herpetologica 28(4):376-380.—Physiology. *Elaphe, Pituophis.*

894 **Landreth, H. F.** 1973. Orientation and behavior of the rattlesnake, *Crotalus atrox.* Copeia 1973(1):26-31.—Behavior, Techniques. *Crotalus.*

895 **Landreth, H. F., and M. T. Christensen.** 1971. Orientation of the plains spadefoot toad, *Scaphiopus bombifrons,* to solar cues. Herpetologica 27(4):454-461.—Behavior. *Scaphiopus.*

896 **Lane, H. H.** 1909. *Alligator mississippiensis* in Oklahoma. Science 30(782):923-924.—Distribution. *Alligator.*

897 **Lane, H. H.** 1909. A paired endoplaston in *Trionyx* and its significance. Proc. Indiana Acad. Sci. 6 pp.—Volume and issue not indicated on article. Development, Morphology. *Trionyx.*

898 **Lane, H. H.** 1926. Oklahoma. Pages 490-502 *in* V. E. Shelford (Ed.). naturalist's guide to the Americas. Williams and Wilkins Co., Baltimore.—Natural History. *Bufo, Rana, Macroclemys, Terrapene, Trionyx, Ophisaurus, Crotaphytus, Phrynosoma, Eumeces, Cnemidophorus, Coluber, Elaphe, Heterodon, Lampropeltis, Pituophis, Leptotyphlops, Agkistrodon, Crotalus, Alligator.*

Langston, W., Jr., *see* 1377

899 **Lardie, G. E., and R. L. Lardie.** 1976. Striped pattern in a western massasauga. Bull. Okla. Herp. Soc. 1(3):40.—Morphology. *Sistrurus.*

900 **Lardie, G. E., and R. L. Lardie.** 1980. Northwestern Oklahoma reptile county records. Bull. Okla. Herp. Soc. 5(1):9.—Distribution. *Kinosternon, Crotaphytus.*

901 **Lardie, G. E., and R. L. Lardie.** 1980. Notes on hatching of eggs of a common snapping turtle after natural underwater deposition. Bull. Okla. Herp. Soc. 4(4):71.—Natural History. *Chelydra.*

902 **Lardie, R. L.** 1973. The amphibians and reptiles of Garfield County. M.S. Thesis, Phillips Univ., Enid, Oklahoma. 94 pp.—Distribution, Ecology, Natural History. All Anura, *Ambystoma, Chelydra, Chrysemys, Pseudemys, Trachemys, Terrapene, Kinosternon, Trionyx, Ophisaurus, Crotaphytus, Holbrookia, Phrynosoma, Sceloporus, Eumeces, Scincella, Cnemidophorus, Arizona, Carphophis, Cemophora, Coluber, Elaphe, Heterodon, Hypsiglena, Lampropeltis, Masticophis, Nerodia, Opheodrys, Pituophis, Regina, Rhinocheilus, Sonora, Storeria, Tantilla, Thamnophis, Tropidoclonion, Virginia, Leptotyphlops, Agkistrodon, Crotalus, Sistrurus.*

903 **Lardie, R. L.** 1975. Courtship and mating in the yellow mud turtle, *Kinosternon flavescens flavescens.* J. Herpetol. 9(2):223-227.—Behavior. *Kinosternon.*

904 **Lardie, R. L.** 1975. Geographic distribution: *Chrysemys floridana hoyi* Herpetol. Rev. 6(4):116.—Distribution. *Chrysemys.*

905 **Lardie, R. L.** 1975. Geographic distribution: *Tantilla nigriceps nigriceps.* Herpetol. Rev. 6(2):45.—Distribution. *Tantilla.*

906 **Lardie, R. L.** 1975. Geographic distribution: *Terrapene carolina triunguis.* Herpetol. Rev. 6(2):44.—Distribution. *Terrapene.*

907 **Lardie, R. L.** 1975. Geographic distribution: *Thamnophis sirtalis annectens.* Herpetol. Rev. 6(4):116.—Distribution. *Thamnophis.*

908 **Lardie, R. L.** 1975. Observations on reproduction in *Kinosternon.* J. Herpetol. 9(2):260-264.—Natural History. *Kinosternon.*

909 **Lardie, R. L.** 1976. Brood of Texas garter snakes from Oklahoma. Bull. Okla. Herp. Soc. 1(3):38.—Distribution. *Thamnophis.*

910 **Lardie, R. L.** 1976. Distributional notes on the checkered garter snake and other *Thamnophis* in north-central Oklahoma. Bull. Okla. Herp. Soc. 1(4):56-57.—Distribution. *Bufo, Thamnophis.*

911 **Lardie, R. L.** 1976. Eggs and young of a Great Plains ground snake from Garfield County, Oklahoma. Bull. Okla. Herp. Soc. 1(4):63-64.—Natural History. *Sonora.*

912 **Lardie, R. L.** 1976. Eggs of the Great Plains ground snake *Sonora episcopa episcopa* from Texas and Oklahoma. Bull. Okla. Herp. Soc. 1(2):19.—Natural History, Techniques. *Sonora.*

913 **Lardie, R. L.** 1976. Geographic distribution: *Cemophora coccinea copei.* Herpetol. Rev. 7(3):124.—Distribution. *Cemophora.*

914 **Lardie, R. L.** 1976. Incubation method of some five-lined skink eggs. Bull. Okla. Herp. Soc. 1(1):2.—Natural History, Techniques. *Eumeces.*

915 **Lardie, R. L.** 1976. Incubation of turtle eggs from a road kill. Bull. Okla. Herp. Soc. 1(2):23.—Natural History, Techniques. *Trachemys.*

916 **Lardie, R. L.** 1976. Injury and treatment of a large prairie kingsnake. Bull. Okla. Herp. Soc. 1(1):4-5.—Techniques. *Coluber, Lampropeltis.*

917 **Lardie, R. L.** 1976. Large centipede eaten by a western massasauga. Bull. Okla. Herp. Soc. 1(3):40.—Natural History. *Sistrurus.*

918 **Lardie, R. L.** 1976. Notes on a rainy night in north-central Oklahoma. Bull. Okla. Herp. Soc. 1(4):69-70.—Natural History. *Scaphiopus, Ambystoma, Pituophis, Thamnophis, Leptotyphlops.*

919 **Lardie, R. L.** 1976. Record-sized brood of checkered garter snakes. Bull. Okla. Herp. Soc. 1(4):67.—Natural History. *Thamnophis.*

920 **Lardie, R. L.** 1977. Herpetology of Garfield County, Oklahoma. Spec. Publ. Okla. Herp. Soc. (1):1-26.—Distribution, Ecology, Natural History. All Anura, *Ambystoma, Chelydra, Macroclemys, Chrysemys, Pseudemys, Trachemys, Terrapene, Kinosternon, Trionyx, Ophisaurus, Crotaphytus, Holbrookia, Phrynosoma, Sceloporus, Eumeces, Scincella, Cnemidophorus, Arizona, Carphophis, Cemophora, Coluber, Elaphe, Heterodon, Hypsiglena, Lampropeltis, Masticophis, Nerodia, Opheodrys, Pituophis, Regina, Rhinocheilus, Sonora, Storeria, Tantilla, Tham-*

nophis, Tropidoclonion, Virginia, Leptotyphlops, Agkistrodon, Crotalus, Sistrurus.

921 **Lardie, R. L.** 1977. States that protect venomous reptiles. Bull. Okla. Herp. Soc. 2(1):19.—General Reference, Management.

922 **Lardie, R. L.** 1978. Additional observations on courtship and mating in the Plains yellow mud turtle, *Kinosternon flavescens flavescens.* Bull. Okla. Herp. Soc. 3(4):70-72.—Behavior. *Kinosternon.*

923 **Lardie, R. L.** 1978. Additional reptile records for Blaine County, Oklahoma. Bull. Okla. Herp. Soc. 3(3):55.—Distribution. *Trachemys, Trionyx.*

924 **Lardie, R. L.** 1978. Longevity record for the eastern hognose snake. Bull. Okla. Herp. Soc. 3(3):57.—Natural History. *Bufo, Scaphiopus, Heterodon.*

925 **Lardie, R. L.** 1979. Eggs and young of the Plains yellow mud turtle. Bull. Okla. Herp. Soc. 4(2-3):24-32.—Natural History. *Kinosternon.*

926 **Lardie, R. L.** 1979. Herpetological records from northwestern Oklahoma. Herpetol. Rev. 10(1):24-25.—Distribution. *Graptemys, Trachemys, Terrapene, Kinosternon, Trionyx, Ophisaurus, Crotaphytus, Holbrookia, Sceloporus, Eumeces, Scincella, Arizona, Cemophora, Coluber, Diadophis, Elaphe, Heterodon, Hypsiglena, Lampropeltis, Masticophis, Nerodia, Opheodrys, Regina, Rhinocheilus, Sonora, Storeria, Thamnophis, Tropidoclonion, Crotalus, Sistrurus.*

927 **Lardie, R. L.** 1980. Reptile county records for northwestern Oklahoma. Bull. Okla. Herp. Soc. 4(4):77.—Distribution. *Graptemys, Arizona, Nerodia.*

928 **Lardie, R. L.** 1980. Some herpetology of the Gypsum Hills of Kingfisher County, Oklahoma, and county verification list. Bull. Okla. Herp. Soc. 5(3):49-53.—Distribution. *Bufo, Acris, Pseudacris, Rana, Ambystoma, Chelydra, Trachemys, Terrapene, Kinosternon, Trionyx, Crotaphytus, Phrynosoma, Cnemidophorus, Arizona, Coluber, Diadophis, Elaphe, Heterodon, Lampropeltis, Masticophis, Nerodia, Pituophis, Regina, Sonora, Thamnophis, Tropidoclonion, Crotalus, Sistrurus.*

929 **Lardie, R. L.** 1981. Collections and observations in Kingfisher County, Oklahoma with new records from the Gypsum Hills. Bull. Okla. Herp. Soc. 6(3):62-64.—Distribution. *Acris, Rana, Chelydra, Trachemys, Trionyx, Crotaphytus, Phrynosoma, Sceloporus, Eumeces, Cnemidophorus, Diadophis, Elaphe, Heterodon, Masticophis, Nerodia, Pituophis, Sonora, Tantilla, Thamnophis, Tropidoclonion, Crotalus.*

930 **Lardie, R. L.** 1981. Some new and important county records for Oklahoma. Bull. Okla. Herp. Soc. 6(2):44-45.—Distribution. *Kinosternon, Sceloporus, Thamnophis, Leptotyphlops.*

931 **Lardie, R. L.** 1982. Feeding fish and beef to snakes. Bull. Okla. Herp. Soc. 7(1):17.—Techniques. *Rana, Heterodon, Nerodia, Thamnophis.*

932 **Lardie, R. L.** 1982. An Oklahomans version of the Chickaskia River Wildlife Study. Bull. Okla. Herp. Soc. 7(3-4):83-85.—Distribution. *Acris, Pseudacris, Gastrophryne, Rana, Chelydra, Chrysemys, Graptemys, Terrapene, Kinosternon, Trionyx, Eumeces, Cnemidophorus, Lampropeltis, Nerodia, Pituophis.*

933 **Lardie, R. L.** 1982. A preliminary checklist of the amphibians and reptiles of northwestern Oklahoma. (Excluding the Oklahoma Panhandle). Bull. Okla. Herp. Soc. 7(2):36-78.—Distribution. All Anura. *Ambystoma, Notophthalmus, Chelydra, Macroclemys, Chrysemys, Graptemys, Trachemys, Terrapene, Kinosternon, Sternotherus, Trionyx, Ophisaurus, Crotaphytus, Holbrookia, Phrynosoma, Sceloporus, Eumeces, Scincella, Cnemidophorus, Arizona, Cemophora, Coluber, Diadophis, Elaphe, Heterodon, Hypsiglena, Lampropeltis, Masticophis, Nerodia, Opheodrys, Pituophis, Regina, Rhinocheilus, Sonora, Storeria, Tantilla, Thamnophis, Tropidoclonion, Virginia, Leptotyphlops, Agkistrodon, Crotalus, Sistrurus.*

934 **Lardie, R. L.** 1983. Aggressive interaction among melanistic males of the red-eared slider, *Pseudemys scripta elegans* (Wied). Bull. Okla. Herp. Soc. 8(4):105-117.—Behavior, Natural History. *Chelydra, Graptemys, Trachemys, Kinosternon, Sternotherus, Trionyx.*

935 **Lardie, R. L.** 1983. Aggressive interactions and territoriality in the yellow mud turtle, *Kinosternon flavescens flavescens* (Agassiz). Bull. Okla. Herp. Soc. 8(3):68-83.—Behavior, Ecology. *Kinosternon.*

936 **Lardie, R. L.** 1983. Melanistic bullsnake from Garfield County, Oklahoma. Bull. Okla. Herp. Soc. 8(3):88-89.—Morphology. *Pituophis.*

937 **Lardie, R. L.** 1983. Transfers, changes, corrections and additions to northwestern Oklahoma herpetological records. Bull. Okla. Herp. Soc. 8(3):84-87.—Distribution. *Pseudacris, Gastrophryne, Ambystoma, Kinosternon, Crotaphytus, Holbrookia, Carphophis, Elaphe, Heterodon, Lampropeltis, Sonora, Tantilla.*

938 **Lardie, R. L.** 1984. A personal opinion. Bull. Okla. Herp. Soc. 9(3):20-21.—General Reference, Popular.

939 **Lardie, R. L.** 1984. Polypropylene tubs as turtle containers. Bull. Okla. Herp. Soc. 9(3):29-30—Techniques. *Chelydra, Deirochelys, Trachemys, Terrapene, Kinosternon, Sternotherus, Trionyx.*

940 **Lardie, R. L.** 1985. Two additional specimens of the timber rattlesnake, *Crotalus horridus* Linnaeus, from north central Oklahoma. Bull. Okla. Herp. Soc. 10(1-2):6-7.—Distribution. *Crotalus.*

941 **Lardie, R. L.** 1985. A second rough green snake, *Opheodrys aestivus* (Linnaeus), recorded in Garfield County, Oklahoma. Bull. Okla. Herp. Soc. 10(1-2):10-11.—Distribution, Natural History. *Opheodrys.*

942 **Lardie, R. L.** 1985. The western distribution pattern of some eastern amphibians and reptiles in Oklahoma. Bull. Okla. Herp. Soc. 10(1-2): 11-12.—Distribution. *Bufo, Hyla, Rana, Ambystoma, Notophthalmus, Graptemys, Pseudemys, Terrapene, Sternotherus, Eumeces, Scincella, Carphophis, Opheodrys, Tantilla, Agkistrodon, Crotalus.*

943 **Lardie, R. L.** 1986. Two distribution records for reptiles in northwestern Oklahoma. Bull. Okla. Herp. Soc. 11(1-4):5.—Distribution. *Graptemys, Hypsiglena.*

944 **Lardie, R. L.** 1987. First record of Graham's crayfish snake, *Regina grahamii* Baird and Girard, from Woods County, Oklahoma. Bull. Okla. Herp. Soc. 12(1-4):19.—Distribution. *Coluber, Regina.*

945 **Lardie, R. L., and J. H. Black.** 1981. The amphibians and reptiles of the Cimarron Gypsum Hills region in northwestern Oklahoma. Bull. Okla. Herp. Soc. 5(4):76-125.—Distribution, Ecology. All Anura. *Ambystoma, Chelydra, Macroclemys, Chrysemys, Graptemys, Pseudemys, Trachemys, Terrapene, Kinosternon, Trionyx, Ophisaurus, Crotaphytus, Holbrookia, Phrynosoma, Sceloporus, Eumeces, Scincella, Cnemidophorus, Arizona, Coluber, Diadophis, Elaphe, Heterodon, Hypsiglena, Lampropeltis, Masticophis, Nerodia, Opheodrys, Pituophis, Regina, Rhinocheilus, Sonora, Storeria, Tantilla, Thamnophis, Tropidoclonion, Virginia, Leptotyphlops, Agkistrodon, Crotalus, Sistrurus.*

946 **Lardie, R. L., and B. P. Glass.** 1975. Geographic distribution: *Ambystoma texanum.* Herpetol. Rev. 6(4):115.—Distribution. *Ambystoma.*

Lardie, R. L., *see* 107, 154, 664, 692, 899, 900, 1056

947 **Larkin, P.** 1910. The occurrence of a sauropod dinosaur in the Trinity Cretaceaus of Oklahoma. J. Geol. 28(1):93-98.—General Reference, Paleontology.

948 **Larson, A.** 1984. Neontological inferences of evolutionary pattern and process in the salamander family Plethodontidae. Pages 119-217 *in* M. K. Hecht, B. Wallace, and G. T. Prance (Eds.). Evolutionary biology, Vol. 17. Plenum Publ. Corp. 351 pp.—Evolution, Genetics. *Plethodon.*

949 **Larson, A., and R. Highton.** 1978. Geographic protein variation and divergence in the salamanders of the *Plethodon welleri* group (Amphibia: Plethodontidae). Syst. Zool. 27(4):431-448.—Distribution, Genetics. *Plethodon.*

Larson, A., *see* 774

Latham, E., *see* 154

950 **Latorre, V.** 1976. Winter hatchling snapping turtle. Bull. Okla. Herp. Soc. 1(2):23.—Natural History. *Chelydra.*

951 **Laubach, H. E., and J. E. Jordan.** 1976. Acanthocephalan parasite of red-eared turtles. Bull. Okla. Herp. Soc. 1(4):66.—Parasitology. *Trachemys.*

952 **Laughlin, H. E.** 1959. Stomach contents of some aquatic snakes from Lake McAlester, Pittsburgh County, Oklahoma. Tex. J. Sci. 11(1):83-85.—Natural History. *Nerodia, Regina, Thamnophis, Agkistrodon.*

953 **Laughlin, H. E.** 1964. Some distributional records of amphibians and reptiles from southeastern Oklahoma. Proc. Okla. Acad. Sci. 44:61-63.—Distribution. *Amphiuma, Eumeces, Diadophis, Heterodon, Virginia.*

Laughlin, H. E., *see* 181, 1399

954 **Leake, D. V.** 1938. Preliminary notes on the production of motile cells of *Basicladia.* Proc. Okla. Acad. Sci. 19:109-110.—Article discusses the life cycle of a commensal alga that grows on turtles. Natural History. *Chelydra.*

955 **Leake, D. V.** 1948. Asexual reproduction in the genus *Basicladia,* Hoffman and Tilden. Proc. Okla. Acad. Sci. 29:17-20.—Article discusses a commensal alga that grows on turtles. Parasitology. *Chelydra, Kinosternon.*

956 **Legler, J. M.** 1960. Natural history of the ornate box turtle, *Terrapene ornata ornata* Agassiz. Univ. Kans. Publ. Mus. Nat. Hist. 11(10):527-669.—Distribution, Natural History. *Terrapene.*

957 **Lemon.** 1959. Okla. Wildl. 15(6):17.—Photograph of a snapping turtle. Incorrectly printed as issue 4. Popular. *Macroclemys.*

Leonard, A. B., *see* 1331

958 **Leuck, B. E., and C. C. Carpenter.** 1981. Shell variation in a population of three-toed box turtles (*Terrapene carolina triunguis*). J. Herpetol. 15(1):53-58.—Morphology. *Terrapene.*

Lhotka, J. F., Jr., *see* 6

959 **Lincoln, K.** 1967. Rattlesnake roundup. Outdoor Okla. 23(6):18-19.—Popular. *Crotalus.*

960 **Lindsay, H. L., Jr.** 1954. The narrow-mouthed toads *Microhyla olivacea* and *M. carolinensis* in northeastern Oklahoma. Copeia 1954(2):158.—Distribution, Natural History. *Hyla, Pseudacris, Gastrophryne.*

961 **Lindsay, H. L., Jr.** 1961. Additional records of *Ambystoma annulatum* from Arkansas and Oklahoma. Southwest. Nat. 6(3-4):199-200.—Distribution. *Ambystoma.*

Lindsay, H. L., Jr., *see* 182, 183

962 **Lintz, C.** 1982. The Wheeler Site (34Pu-102), phase II, archeological investigations of Clayton Lake, southeast Oklahoma. Archeol. Res. Manage. Center Univ. Okla. Res. Ser. 8:189-253.—Turtle shell found. Archeology. *Terrapene.*

963 **Lish, J. W.** 1973. Bald eagles wintering on the Neosho River, Oklahoma. Bull. Okla. Ornithol. Soc. 6(4):23-30.—Snapping turtle remains were found under eagle perch. Natural History. *Chelydra.*

964 **Lish, J. W.** 1973. Status and ecology of bald eagles and nesting of golden eagles in Oklahoma. M.S. Thesis, Okla. State Univ., Stillwater. 98 pp.—Plastron and carapace of ornate box turtle found in golden eagle eyries. Natural History. *Terrapene.*

965 **Lish, J. W., and S. K. Sherrod.** 1986. A history of bald eagle nesting activity in Oklahoma. Proc. Okla. Acad. Sci. 66:15-20.—Soft-shelled turtle remains were found under nest. Natural History. *Trionyx.*

966 **List, J. S.** 1966. Comparative osteology of the snake families Typhlopidae and Leptotyphlopidae. Illinois Biol. Monogr. (36):1-112.—Evolution, Morphology, Taxonomy. *Leptotyphlops.*

967 **Littlejohn, M. J.** 1960. Call discrimination and potential reproductive isolation in *Pseudacris triseriata* females from Oklahoma. Copeia 1960(4):370-371.—Behavior. *Pseudacris.*

968 **Littlejohn, M. J., M. J. Fouquette, Jr., and C. Johnson.** 1960. Call discrimination by female frogs of the *Hyla versicolor* complex. Copeia 1960(1):47-49.—Behavior. *Hyla.*

969 **Littlejohn, M. J., and R. S. Oldham.** 1968. *Rana pipiens* complex: Mating call structure and taxonomy. Science 162(3857):1003-1005.—Behavior, Distribution, Taxonomy. *Rana.*

970 **Litvaitis, J. A.** 1975. Movements and habitat use of coyotes on the Wichita Mountains National Wildlife Refuge. M.S. Thesis, Okla. State Univ., Stillwater. 69 pp.—Unidentified reptiles found in 13 of 361 scats. General Reference, Natural History.

Lohoefener, R., *see* 9

Lollis, D. W., *see* 155

971 **Loomis, R. B.** 1955. *Trombicula gurneyi* Ewing and two new related chigger mites (Acarina, Trombiculidae). Univ. Kans. Sci. Bull. 37(9): 251-267.—Parasitology. *Sceloporus, Eumeces.*

972 **Loomis, R. B., and O. L. Webb.** 1951. *Eurycea multiplicata* collected at the restricted type locality. Herpetologica 7(3):141-142.—Distribution, Morphology, Natural History. *Eurycea.*

Loomis, R. B., *see* 1390

973 **Loughry, W. J.** 1987. The dynamics of snake harassment by black-tailed prairie dogs. Behaviour 103(1-3):27-48.—Prairie rattlesnakes from Oklahoma rattlesnake roundups were used. Behavior. *Crotalus.*

Loughry, W. J., *see* 1163

974 **Loving, D.** 1978. New county records for the pallid softshell turtle. Bull. Okla. Herp. Soc. 3(3):55.—Distribution. *Trionyx.*

975 **Loving, D., J. H. Black, M. Butcher, and E. M. Palmer.** 1978. Mississippi mud turtle in Murray County, Oklahoma. Bull. Okla. Herp. Soc. 3(3):56.—Distribution. *Kinosternon.*

Loving, D., *see* 150

976 **Lowe, C. H., Jr.** 1966. The prairie lined racerunner. J. Ariz. Acad. Sci. 4(1):44-45.—Taxonomy. *Cnemidophorus.*

977 **Lund, R.** 1978. Anatomy and relationships of the family Phlegethontiidae (Amphibia, Aistopoda). Ann. Carnegie Mus. 47(4):53-79.—General Reference, Paleontology.

978 **Lynch, J. E.** 1933. The miracedium of *Hermonimus chelydrae* MacCallum. Q.J. Micosc. Sci. 76(1):13-33.—Parasitology. *Chelydra.*

979 **Lynn, R. T.** 1963. Comparative behavior of the horned lizards, genus *Phrynosoma,* of the United States. Ph.D. Diss., Univ. Okla., Norman. 76 pp.—Behavior. *Phrynosoma.*

980 **Lynn, R. T.** 1965. A comparative study of the display behavior in *Phrynosoma* (Iguanidae). Southwest. Nat. 10(1):25-30.—Behavior. *Phrynosoma.*

981 **Mahaffey, J.** 1945. Poisonous pleasure. Okla. Game Fish News 1(6): 10-11, 19.—Popular. *Rana, Crotalus.*

982 **Mahaffey, J.** 1946. Beaver's Bend State Park. Okla. Game Fish News 2(7):4, 6-7.—Lists some snakes in the park. Popular. *Agkistrodon, Crotalus.*

983 **Mahaffey, J.** 1954. Snakes alive. Okla. Game Fish News 10(9):10-11.—Natural History, Popular. *Coluber, Heterodon, Agkistrodon, Sistrurus.*

984 **Mahaffey, J.** 1957. Our friends: The frogs and toads. Okla. Game Fish News 13(9):12-13.—Popular. *Bufo, Hyla, Rana, Ambystoma.*

985 **Mahaffey, J.** 1958. May time. Okla. Wildl. 14(5):12-13.—Popular. *Terrapene, Phrynosoma, Eumeces.*

Mahan, J. T., *see* 98

986 **Mahmoud, I. Y.** 1960. The comparative ecology of the kinosternid turtles of Oklahoma. Ph.D. Diss., Univ. Okla., Norman. 229 pp.—Behavior, Ecology, Natural History. *Kinosternon, Sternotherus.*

987 **Mahmoud, I. Y.** 1967. Courtship behavior and sexual maturity in four species of kinosternid turtles. Copeia 1967(2):314-318.—Behavior. *Kinosternon, Sternotherus.*

988 **Mahmoud, I. Y.** 1968. Feeding behavior in kinosternid turtles. Herpetologica 24(4):300-305.—Ecology. *Kinosternon, Sternotherus.*

989 **Mahmoud, I. Y.** 1969. Comparative ecology of the kinosternid turtles of Oklahoma. Southwest. Nat. 14(1):31-66.—Behavior, Ecology, Techniques. *Kinosternon, Sternotherus.*

990 **Mahmoud, I. Y., and J. Klicka.** 1972. Seasonal gonadal changes in kinosternid turtles. J. Herpetol. 6(3-4):183-189.—Physiology. *Kinosternon, Sternotherus.*

991 **Marcy, R. B.** 1853. Exploration of the Red River of Louisiana in the year 1852. Senate Exec. Doc. No. 54, 32nd Cong. 2nd Sess. 320 pp.—Appendix F contains "reptiles" by S. F. Baird and C. Girard. Distribution, Historical. *Bufo, Rana, Crotaphytus, Holbrookia, Phrynosoma, Sceloporus, Scincella, Cnemidophorus, Elaphe, Heterodon, Lampropeltis, Masticophis, Opheodrys, Pituophis, Thamnophis, Crotalus.*

992 **Marr, J. C.** 1944. Notes on amphibians and reptiles from the central United States. Am. Midl. Nat. 32(2):478-490.—Distribution. *Bufo, Acris, Rana, Terrapene, Kinosternon, Crotaphytus, Holbrookia, Phrynosoma, Sceloporus, Eumeces, Arizona, Heterodon, Masticophis, Thamnophis, Crotalus.*

993 **Martin, P. S., and B. E. Harrell.** 1957. The Pleistocene history of temperate biotas in Mexico and eastern United States. Ecology 38(3): 468-480.—Paleontology. *Storeria, Sistrurus.*

Martin, R., *see* 106

994 **Martin, R. F.** 1972. Evolution of vocalization in the genus *Bufo*. Pages 279-309 *in* W. F. Blair (Ed.). Evolution in the genus *Bufo*. Univ. Tex. Press, Austin. 459 pp.—Behavior. *Bufo.*

995 **Martin, R. F.** 1973. Osteology of North American *Bufo:* the *americanus, cognatus,* and *boreas* species groups. Herpetologica 29(4): 375-387.—Morphology. *Bufo.*

996 **Martof, B. S.** 1973. *Siren intermedia.* Cat. Am. Amphib. Reptiles 127.1-127.3.—Range map and literature review. Bibliographies, Distribution. *Siren.*

997 **Martof, B. S.** 1974. Sirenidae. Cat. Am. Amphib. Reptiles 151.1-151.2.—Range map and literature review. Bibliographies, Distribution. *Siren.*

998 **Martof, B. S.** 1974. *Siren.* Cat. Am. Amphib. Reptiles 152.1-152.2.—Range map and literature review. Bibliographies, Distribution. *Siren*

999 **Martof, B. S.** 1975. *Hyla squirella.* Cat. Am. Amphib. Reptiles 168.1-168.2.—A species thought to occur in Oklahoma, but not confirmed. Range map and literature review. Bibliographies, Distribution. *Hyla.*

1000 **Maslin, T. P.** 1953. The status of the whipsnake *Masticophis flagellum* (Shaw) in Colorado. Herpetologica 9(4):193-200.—Distribution, Taxonomy. *Masticophis.*

1001 **Maslin, T. P.** 1956. *Sceloporus undulatus erythrocheilus* ssp. nov. (Reptilia, Iguanidae) from Colorado. Herpetologica 12(4):291-294.—Distribution, Taxonomy. *Sceloporus.*

Maslin, T. P., *see* 640

Matalas, B. L., *see* 753

1002 **Mather, C. M.** 1978. Geographic distribution: *Rhinocheilus lecontei tesselatus.* Herpetol. Rev. 9(4):142-143.—Distribution. *Rhinocheilus.*

1003 **Mather, C. M.** 1986. New county records for reptiles in Grady County, Oklahoma. Bull. Okla. Herp. Soc. 11(1-4):14.—Distribution. *Chelydra, Coluber, Heterodon, Nerodia, Tropidoclonion, Virginia, Agkistrodon.*

Mathews, R., *see* 375

1004 **Maxwell, J.** 1970. Sounds of spring. Outdoor Okla. 26(4):17-19.—Popular. *Bufo, Acris, Hyla, Pseudacris, Rana.*

1005 **Mays, M. L.** 1960. A survey of helminth parasites of box turtles in central Oklahoma. M.S. Thesis, Univ. Okla., Norman. 23 pp.—Parasitology. *Terrapene.*

1006 **McAllister, C. T.** 1985. *Nerodia rhombifera.* Cat. Am. Amphib. Reptiles 376.1-376.4—Range map and literature review. Bibliographies, Distribution. *Nerodia.*

1007 **McCardell, J. H.** 1971. The influence of temperature on the duration of the aggressive display of *Anolis carolinensis.* M.S. Thesis, Univ. Okla., Norman. 39 pp.—Behavior. *Anolis.*

1008 **McCarley, H. W.** 1953. New locality records for two Oklahoma salamanders. Copeia 1953(1):62-63.—Distribution. *Siren, Notophthalmus.*

1009 **McCarley, H. W.** 1970. *Rana areolata* in southern Oklahoma-northern Texas. Southwest. Nat. 15(2)261-273.—Distribution. *Rana.*

McCarley, H. W., *see* 224

1010 **McClung, D. J.** 1980. Winter activity of *Chrysemys scripta elegans* (Wied-Neuwied) in Oklahoma. Bull. Okla. Herp. Soc. 4(4):70.—Natural History. *Trachemys.*

1011 **McConkey, E. H.** 1951. A systematic study of the North American lizards of the genus *Ophisaurus.* M.S. Thesis, Univ. Florida, Gainesville. 107 pp.—Distribution, Ecology, Natural History, Taxonomy. *Ophisaurus.*

1012 **McConkey, E. H.** 1952. A new subspecies of *Ophisaurus attenuatus,* with a key to the North American forms. Nat. Hist. Misc. (Chic.) (102):2-3.—Distribution. *Ophisaurus.*

1013 **McConkey, E. H.** 1954. A systematic study of the North American lizards of the genus *Ophisaurus.* Am. Midl. Nat. 51(1):133-171.—Distribution, Taxonomy. *Ophisaurus.*

1014 **McCoy, C. J., Jr.** 1958. Daily activity of some Oklahoma lizards. J. Ohio Herpetol. Soc. 1(3):12.—Natural History. *Crotaphytus, Holbrookia, Phrynosoma, Sceloporus, Eumeces, Cnemidophorus.*

1015 **McCoy, C. J., Jr.** 1959. New county records and range extensions for Oklahoma reptiles. Proc. Okla. Acad. Sci. 40:41-43.—Distribution. *Chelydra, Macroclemys, Pseudemys, Terrapene, Kinosternon, Sternotherus, Trionyx, Ophisaurus, Eumeces, Carphophis, Coluber, Diadophis, Elaphe, Heterodon, Lampropeltis, Masticophis, Nerodia, Opheodrys, Regina, Rhinocheilus, Sonora, Storeria, Tantilla, Thamnophis, Tropidoclonion, Virginia, Leptotyphlops, Agkistrodon, Sistrurus.*

1016 **McCoy, C. J., Jr.** 1960. Systematics and variation of *Sceloporus undulatus* in Oklahoma. M.S. Thesis, Okla. State Univ., Stillwater. 55 pp.—Distribution, Taxonomy. *Sceloporus.*

1017 **McCoy, C. J., Jr.** 1960. An unusually large aggregation of *Leptotyphlops.* Copeia 1960(4):368.—Thirteen snakes were found under a rock north of Stillwater. Ecology. *Leptotyphlops.*

1018 **McCoy, C. J., Jr.** 1961. Birth season and young of *Crotalus scutulatus* and *Agkistrodon contortrix laticinctus.* Herpetologica 17(2):140.—Natural History. *Agkistrodon.*

1019 **McCoy, C. J., Jr.** 1961. Distribution of the subspecies of *Sceloporus undulatus* (Reptilia: Iguanidae) in Oklahoma. Southwest. Nat. 6(2): 79-85.—Distribution. *Sceloporus.*

1020 **McCoy, C. J., Jr.** 1961. A technique for collecting in urban areas. J. Ohio Herpetol. Soc. 3(1):4.—This is of author's experiences while in Oklahoma. Techniques. *Pseudacris, Gastrophryne, Storeria, Tropidoclonion.*

1021 **McCoy, C. J., Jr.** 1961. *Thamnophis marcianus* in central Oklahoma. J. Ohio Herpetol. Soc. 3(2):23-24.—Distribution. *Thamnophis.*

1022 **McCoy, C. J., Jr.** 1966. The development of melanism in an Oklahoma population of *Chrysemys scripta elegans* (Reptilia: Testudinidae). Proc. Okla. Acad. Sci. 47:84-87.—Management. *Trachemys.*

1023 **McCoy, C. J., Jr.** 1967. Diet of bullfrogs (*Rana catesbeiana*) in central Oklahoma farm ponds. Proc. Okla. Acad. Sci. 48:44-45.—Natural History. *Bufo, Hyla, Rana, Terrapene, Nerodia.*

1024 **McCoy, C. J., Jr.** 1970. *Hemidactylus turcicus.* Cat. Am. Amphib. Reptiles 87.1-87.2.—This provides a literature review of an introduced species now established in the state. Distribution, General Reference.

1025 **McCoy, C. J., Jr.** 1975. Cave-associated snakes, *Elaphe guttata,* in Oklahoma. N. S. S. Bull. 37(2):41.—Ecology. *Elaphe.*

1026 **McCoy, C. J., Jr.** 1975. Notes on Oklahoma reptiles. Proc. Okla. Acad. Sci. 55:53-54.—Distribution. *Chrysemys, Kinosternon, Trionyx, Sceloporus, Scincella, Arizona, Elaphe, Storeria, Virginia, Sistrurus.*

1027 **McCoy, C. J., Jr.** 1976. Geographic distribution: *Storeria dekayi texana.* Herpetol. Rev. 7(3):123.—Distribution. *Storeria.*

1028 **McCoy, C. J., Jr.** 1976. Geographic distribution: *Thamnophis radix haydeni.* Herpetol. Rev. 7(3):124.—Distribution. *Thamnophis.*

1029 **McCoy, C. J., Jr.** 1976. Geographic distribution: *Tropidoclonion lineatum.* Herpetol. Rev. 7(3):124.—Distribution. *Tropidoclonion.*

1030 **McCoy, C. J., Jr.** 1977. Notes on Oklahoma reptiles. Bull. Okla. Herp. Soc. 2(1):11-12.—Distribution. *Trachemys, Kinosternon, Trionyx, Sceloporus, Scincella, Arizona, Virginia, Sistrurus.*

1031 **McCoy, C. J., Jr., and J. H. Black.** 1974. Arthur Norris Bragg (1897-1968): Biography and bibliography. Smithson. Herpetol. Infor. Serv. (27):1-6.—Bibliographies, Historical. *Bufo, Hyla, Scaphiopus, Rana.*

McCoy, C. J., Jr., *see* 1204, 1309, 1456

1032 **McCracken, W. A.** 1966. The reptiles of Woods County, Oklahoma. M.S. Thesis, Okla. State Univ., Stillwater. 34 pp.—Distribution, General Reference.

1033 **McDaniel, V. R., and J. P. Karges.** 1983. *Farancia abacura.* Cat. Am. Amphib. Reptiles 314.1-314.2.—Range map and literature review. Bibliographies, Distribution. *Farancia.*

McGee, L., *see* 1445

1034 **McGinnis, H. J.** 1967. The osteology of *Phlegethontia,* a Carboniferous and Permian aistopod amphibian. Univ. Calif. Publ. Geol. Sci. 71:1-49.—General Reference, Paleontology.

1035 **McGrew, III, W. C.** 1963. Channel catfish feeding on diamond-backed water snakes. Copeia 1963(1):178-179.—Eight juvenile water snakes were found in the stomach of a catfish caught in Lake Eufaula. Natural History. *Nerodia.*

1036 **McKnight, T. J.** 1959. A taxonomic study of the helminth parasites of the turtles of Lake Texoma. Ph.D. Diss., Univ. Okla., Norman.—Library copy missing. General Reference, Parasitology.

1037 **McLain, R. B.** 1899. Notes on a collection of reptiles: Made by Mr. C. J. Pierson at Fort Smith, Arkansas, with remarks on other eastern reptiles. Contrib. N. Am. Herpetol. 5 pp.—Privately printed in Wheeling, West Virginia. Oklahoma included in surrounding area. Historical. *Macroclemys, Trionyx, Ophisaurus, Crotaphytus, Eumeces, Cnemidophorus, Carphophis, Coluber, Heterodon, Lampropeltis, Masticophis, Nerodia, Opheodrys, Tantilla, Thamnophis, Virginia, Agkistrodon.*

1038 **McMullen, D. B.** 1939. Cutaneous myiasis in a box turtle. Proc. Okla. Acad. Sci. 20:23-25.—Parasitology. *Terrapene.*

1039 **McMurry, F. B.** 1942. Amphibians found on the Wichita Mountains Wildlife Refuge.—Mimeographed handout distributed on the refuge by the Fish and Wildlife Service. Distribution. *Bufo, Acris, Pseudacris, Gastrophryne, Scaphiopus, Rana, Ambystoma.*

McMurry, F. B., *see* 1296

1040 **Means, D. B.** 1974. The status of *Desmognathus brimleyorum* Stejneger and an analysis of the genus *Desmognathus* (Amphibia: Urodela) in Florida. Bull. Fla. State Mus. Biol. Sci. 18(1):1-100.—Distribution, Evolution, Morphology, Taxonomy. *Desmognathus.*

1041 **Mecham, J. S.** 1954. Geographic variation in the green frog, *Rana clamitans* Latreille. Tex. J. Sci. 6(1):1-25.—Distribution, Morphology, Taxonomy. *Rana.*

1042 **Mecham, J. S.** 1961. Isolating mechanisms in anuran amphibians. Pages 24-61 *in* W. F. Blair (Ed.). Vertebrate speciation. 642 pp.—Behavior, Genetics, Taxonomy. *Pseudacris.*

1043 **Mecham, J. S.** 1967. *Notophthalmus viridescens.* Cat. Am. Amphib. Reptiles 53.1-53.4.—Range map and literature review. Bibliographies, Distribution. *Notophthalmus.*

1044 **Medica, P. A.** 1975. *Rhinocheilus, R. lecontei.* Cat. Am. Amphib. Reptiles 175.1-175.4.—Range map and literature review. Bibliographies, Distribution. *Rhinocheilus.*

1045 **Medica, P. A.** 1980. Locality records of *Rhinocheilus lecontei* in the United States and Mexico. Herpetol. Rev. 11(2):42.—Distribution. *Rhinocheilus.*

1046 **Mehl, M. G.** 1921. A new form of *Diplocaulus.* J. Geol. 29(1):48-56.—General Reference, Paleontology.

1047 **Mehl, M. G.** 1926. *Trematops thomasi,* a new amphibian species from the Permian of Oklahoma. J. Geol. 34(5):466-474.—General Reference, Paleontology.

1048 **Mellish, C. H.** 1934. Spermatogenesis of the horned lizard, *Phrynosoma cornutum* (Harlan). M.S. Thesis, Univ. Okla., Norman. 43 pp. —Development, Morphology, Techniques. *Phrynosoma.*

Mengden, G. A., *see* 84

1049 **Mensinger, G.** 1973. Trotline-jumpline. Outdoor Okla. 29(4):2-4.—Photograph included of turtle caught on the line.

1050 **Merkle, D. A., and S. I. Guttman.** 1977. Geographic variation in the cave salamander *Eurycea lucifuga.* Herpetologica 33(3):313-321.—Genetics, Taxonomy. *Eurycea.*

Merriman, H. J., *see* 478

1051 **Michaud, T. C.** 1962. Call discrimination by females of the chorus frogs, *Pseudacris clarki,* and *Pseudacris nigrita.* Copeia 1962(1):213-215.—Some of the experimental animals were collected in Oklahoma. Behavior. *Pseudacris.*

Miller, K., *see* 817

1052 **Miller, L.** 1976. Pesticides kill many animals. Bull. Okla. Herp. Soc. 1(3):36.—Countless amphibians are reportedly killed by pesticide misuse. Ecology, General Reference, Health, Management.

1053 **Miller, L.** 1976. Western painted turtles found in Oklahoma. Kans. Herpetol. Soc. Newsl. (15):6.—Distribution. *Chrysemys.*

1054 **Miller, L.** 1977. More western painted turtles found in Oklahoma. Bull. Okla. Herp. Soc. 2(2-3):34.—Distribution. *Chrysemys.*

1055 **Miller, L.** 1979. Life in Kansas and Oklahoma once again threatened by pesticides. Kans. Herpetol. Soc. Newsl. (31):7.—Mentions turtles, snakes, and other reptiles that died. Ecology, Health. *Bufo, Rana.*

1056 **Miller, L., and R. L. Lardie.** 1976. Geographic distribution: *Chrysemys picta belli.* Herpetol. Rev. 7(4):178.—Distribution. *Chrysemys.*

Miller, L. S., *see* 628

Miller, R. L., *see* 1137

1057 **Milstead, W. W.** 1967. Fossil box turtles (*Terrapene*) from central North America, and box turtles of eastern Mexico. Copeia 1967(1):-168-179.—Distribution, Paleontology, Taxonomy. *Terrapene.*

1058 **Milstead, W. W.** 1969. Studies on the evolution of box turtles (Genus *Terrapene*). Bull. Fla. State Mus. Biol. Sci. 14(1):1-113.—Distribution, Evolution, Taxonomy. *Terrapene.*

Milstead, W. W., *see* 71

1059 **Minter, S.** 1979. Geographical distribution: *Ambystoma annulatum.* Herpetol. Rev. 10(1):23.—Distribution. *Ambystoma.*

1060 **Minton, S. A.** 1983. *Sistrurus catenatus.* Cat. Am. Amphib. Reptiles 332.1-332.2.—Range map and literature review. Bibliographies, Distribution. *Sistrurus.*

1061 **Minton, S. A.** 1987. Poisonous snakes and snake bites in the U.S.: A brief review. Northwest Science 61(2):130-137—Health. *Crotalus.*

1062 **Minton, S. A., and S. A. Weinstein.** 1986. Geographic and ontogenetic variation in venom of the western diamondback rattlesnake (*Crotalus atrox*). Toxicon 24(1):71-80.—Health. *Crotalus.*

1063 **Mitchell, J. C.** 1982. *Farancia.* Cat. Am. Amphib. Reptiles 292.1-292.2.—Range map and literature review. Bibliographies, Distribution. *Farancia.*

1064 **Mittleman, M. B.** 1947. American caudata. I. Geographic variation in *Manculus quadridigitatus.* Herpetologica 3(6):209-224.—Included Oklahoma specimens. Distribution. *Eurycea.*

1065 **Mittleman, M. B.** 1949. Geographic variation in Marcy's garter snake, *Thamnophis marcianus* (Baird and Girard). Bull. Chic. Acad. Sci. 8(10):235-249.—Distribution, Morphology, Taxonomy. *Thamnophis.*

1066 **Mittleman, M. B.** 1967. *Manculus.* Cat. Am. Amphib. Reptiles 44.1-44.2.—Range map and literature review. Bibliographies, Distribution. *Eurycea.*

1067 **Monk, S. M., and C. R. Falk.** 1982. A preliminary consideration of vertebrate remains from the Edwards I Site (34BK-2), Beckham County, Oklahoma. Univ. Nebr. Div. Archeol. Res. Tech. Rept. 82-

03.—Carapace and plastron of turtle found. Archeology, General Reference.

1068 **Moodie, R. L.** 1911. The temnospondylous Amphibia and a new species of *Eryops* from the Permian of Oklahoma. Kans. Univ. Sci. Bull. 5(3):235-253.—General Reference, Paleontology.

1069 **Mook, C. C.** 1964. New species of *Goniopholis* from the Morrison of Oklahoma. Okla. Geol. Notes 24(12):283-287.—General Reference, Paleontology.

1070 **Moore, G. A.** 1937. The spadefoot toad under drought conditions. Copeia 1937(4):225-227.—Natural History. *Scaphiopus.*

1071 **Moore, G. A., and E. Byers.** 1938. A case of hermaphroditism in *Rana catesbeiana* Shaw. Trans. Am. Microsc. Soc. 52(4):407-411.—Morphology. *Rana.*

1072 **Moore, G. A., and W. J. Carter.** 1942. A new record for *Eurycea lucifuga* in Oklahoma. Copeia 1942(1):52.—Distribution. *Eurycea.*

1073 **Moore, G. A., and R. C. Hughes.** 1939. A new plethodontid from eastern Oklahoma. Am. Midl. Nat. 22(3):696-699.—Behavior, Distribution, Morphology, Natural History. *Eurycea.*

1074 **Moore, G. A., and R. C. Hughes.** 1941. A new plethodont salamander from Oklahoma. Copeia 1941(3):139-142.—Distribution, Taxonomy. *Eurycea.*

1075 **Moore, G. A., and C. C. Rigney.** 1941. Notes on the herpetology of Payne County, Oklahoma. Proc. Okla. Acad. Sci. 22:77-80.—Distribution. All Anura, *Ambystoma, Chelydra, Graptemys, Pseudemys, Trachemys, Terrapene, Kinosternon, Sternotherus, Trionyx, Ophisaurus, Crotaphytus, Phrynosoma, Sceloporus, Eumeces, Scincella, Cnemidophorus, Coluber, Diadophis, Elaphe, Heterodon, Lampropeltis, Masticophis, Nerodia, Opheodrys, Pituophis, Regina, Sonora, Storeria, Tantilla, Thamnophis, Tropidoclonion, Virginia, Leptotyphlops, Agkistrodon, Crotalus, Sistrurus.*

Moore, G. A., *see* 799

1076 **Moore, J. A.** 1944. Geographic variation in *Rana pipiens* Schreber of eastern North America. Bull. Am. Mus. Nat. Hist. 82(8):345-370.—Morphology, Taxonomy. *Rana.*

1077 **Moore, J. A.** 1946. Incipient intraspecific isolating mechanisms in *Rana pipiens.* Genetics 31(May):304-326.—Development, Genetics. *Rana.*

1078 **Moore, J. A.** 1967. Diploid and haploid hybridization of different populations of the *Rana pipiens* complex II. J. Exp. Zool. 165(3):461-474.—Experiments with frogs from Oklahoma. Development, Genetics. *Rana.*

1079 **Moore, J. A.** 1967. Diploid and haploid hybridization of different populations of the *Rana pipiens* complex I. Experiments with females from Mexico. J. Exp. Zool. 165(1):1-20.—Development, Genetics. *Rana.*

1080 **Moore, J. A.** 1975. *Rana pipiens*-the changing paradigm. Am. Zool. 15(4):837-849.—Distribution, Genetics, Taxonomy. *Rana.*

1081 **Morrison, E. O.** 1967. Mite and lungworm infections in *Acris crepitans* from southern Oklahoma. Tex. J. Sci. 19:328-329.—Parasitology. *Acris.*

1082 **Morrison, E. O.** 1967. *Rhabdias ranae* infections in three species of Oklahoma Anura. Southwest. Nat. 12(3):335-336.—Parasitology. *Acris, Rana.*

1083 **Morton, D. M.** 1982. The behavior of the western slender glass lizard, *Ophisaurus attenuatus attenuatus.* M.S. Thesis, Univ. Okla., Norman. 127 pp.—Behavior. *Ophisaurus.*

1084 **Mosley, K. T., Jr.** 1963. Behavior patterns of the collared lizard (*Crotaphytus collaris collaris*). M.S. Thesis, Univ. Okla., Norman. 50 pp.—Behavior. *Crotaphytus.*

1085 **Muchmore, H. G., A. I. Ortenburger, and A. A. Hellbaum.** 1951. Theraputic conference. J. Okla. State Med. Assoc. 44(8):319-325.—Conference on snake bite treatment. Popular, Techniques. *Agkistrodon, Crotalus.*

Mulvany, J. E., *see* 1096

Mulvany, K. A., *see* 1099

Mulvany, M. C., *see* 1097, 1098, 1099

1086 **Mulvany, P. S.** 1983. Blind snakes of the United States, their natural history with a discussion of climate and physiography as limiting factors to their range. Bull. Okla. Herp. Soc. 8(1):2-45.—Natural History, Taxonomy. *Leptotyphlops.*

1087 **Mulvany, P. S.** 1983. Tarantula preys on western ribbon snake with notes on the relationship between narrowmouth toads and tarantulas. Bull. Okla. Herp. Soc. 8(4):92-99.—Natural History. *Gastrophryne, Diadophis, Thamnophis, Virginia.*

1088 **Mulvany, P. S.** 1984. Curious death of a night snake. Bull. Okla. Herp. Soc. 9(3):31-32.—Feeding methods to avoid death of captive snakes. Natural History, Techniques. *Hypsiglena, Thamnophis, Leptotyphlops.*

1089 **Mulvany, P. S.** 1984. An examination of scale ornament in the anal region of a Texas night snake, *Hypsiglena torquata jani* (Duges), from Stephens County, Oklahoma. Bull. Okla. Herp. Soc. 9(1):4-8.—Morphology. *Hypsiglena.*

1090 **Mulvany, P. S.** 1984. First recordings of *Heterodon* and *Leptotyphlops* in Dewey County, Oklahoma. Bull. Okla. Herp. Soc. 9(2):18.—Distribution. *Heterodon, Leptotyphlops.*

1091 **Mulvany, P. S.** 1984. Ground skink impaled on cottonwood bud. Bull. Okla. Herp. Soc. 9(1):12-13.—Natural History. *Scincella.*

1092 **Mulvany, P. S.** 1984. An instance of early mating and other notes on reproduction in *Tantilla gracilis* Baird and Girard from Tulsa County, Oklahoma. Bull. Okla. Herp. Soc. 9(4):36-38.—Natural History, Techniques. *Tantilla.*

1093 **Mulvany, P. S.** 1985. Simple metal clips for fastening lids to plastic shoe and sweater box cages. Bull. Okla. Herp. Soc. 10(3-4):24-27.—Discussed methods for containing live animals. General Reference, Techniques.

1094 **Mulvany, P. S.** 1986. How the collared lizard became Oklahoma's state reptile. Bull. Okla. Herp. Soc. 11(1-4):20-21.—Historical, Popular. *Crotaphytus.*

1095 **Mulvany, P. S.** 1986. Tongue ecdysis in snakes. Bull. Okla. Herp. Soc. 11(1-4):34-38.—Behavior, Morphology. *Elaphe, Lampropeltis, Pituophis.*

1096 **Mulvany, P. S., and J. E. Mulvany.** 1986. Six pickerel frogs in total darkness in a cave dome pit, Delaware County, Oklahoma. Bull. Okla. Herp. Soc. 11(1-4):15.—Behavior, Ecology. *Rana.*

1097 **Mulvany, P. S., and M. C. Mulvany.** 1984. Seven new county records for snakes in Oklahoma. Bull. Okla. Herp. Soc. 9(2):17.—Distribution. *Coluber, Elaphe, Lampropeltis, Tantilla, Thamnophis, Tropidoclonion, Virginia.*

1098 **Mulvany, P. S., and M. C. Mulvany.** 1985. New county records for reptiles in Craig and Nowata counties, Oklahoma. Bull. Okla. Herp. Soc. 10(1-2):12-14.—Distribution. *Trachemys, Terrapene, Kinosternon,*

Sternotherus, Eumeces, Scincella, Cnemidophorus, Carphophis, Coluber, Diadophis, Lampropeltis, Masticophis, Nerodia, Thamnophis, Virginia, Agkistrodon.

1099 **Mulvany, P. S., M. C. Mulvany, and K. A. Mulvany.** 1987 New records for reptiles in Ellis and Woodward counties, Oklahoma. Bull. Okla. Herp. Soc. 12(1-4):21.—Distribution. *Crotaphytus, Coluber, Elaphe, Lampropeltis, Nerodia, Tantilla, Crotalus.*

1100 **Munn, M.** 1914. Reconnaissance of the Grand Field District, Oklahoma. U.S. Geol. Surv. Bull. (547):1-85.—General Reference, Paleontology.

1101 **Murphy, J.** 1984. Springtime serenade. Outdoor Okla. 40(2):14-19.—Popular. All Anura.

1102 **Murphy, J.** 1985. Frog on the line. Outdoor Okla. 41(4):26-31.—An article on fishing for bullfrogs. Popular. *Rana.*

1103 **Murphy, J.** 1987. Oklahoma lizards. Outdoor Okla. 43(3):30-35.—Popular. *Ophisaurus, Anolis, Crotaphytus, Phrynosoma, Sceloporus, Eumeces, Scincella, Cnemidophorus, Alligator.*

1104 **Murphy, M. F.** 1948. The ecology of the osage wood rat, *Neotoma floridana osagensis* Blair, with emphasis on the helminths of the host. M.S. Thesis, Okla. A.M. Coll., Stillwater. 50 pp.—Water snakes and lizards are included as inhabitants of woodrat dens. Natural History. *Nerodia.*

1105 **Mushinsky, H. R.** 1975. Selection of substrate pH by salamanders. Am. Midl. Nat. 92(2):440-443.—Behavior, Ecology, Techniques. *Eurycea.*

1106 **Neill, W. T.** 1951. The taxonomy of North American soft-shelled turtles, genus *Amyda.* Publ. Res. Div. Ross Allens's Reptile Inst. 1(2): 7-24.—Distribution, Taxonomy. *Trionyx.*

1107 **Neitman, K.** 1977. Dealing with problem feeders in snakes. Bull. Okla. Herp. Soc. 2(2-3):37-38.—Techniques. *Heterodon.*

1108 **Nelson, C. E.** 1972. *Gastrophryne carolinensis.* Cat. Am. Amphib. Reptiles 120.1-120.4.—Range map and literature review. Bibliographies, Distribution. *Gastrophryne.*

1109 **Nelson, C. E.** 1972. *Gastrophryne olivacea.* Cat. Am. Amphib. Reptiles 122.1-122.4—Range map and literature review. Bibliographies, Distribution. *Gastrophryne.*

1110 **Nelson, C. E.** 1972. Systematic studies of the North American microhylid genus *Gastrophryne*. J. Herpetol. 6(2):111-137.—Bibliographies, Ecology, Natural History, Taxonomy. *Gastrophryne*.

1111 **Nelson, C. E.** 1973. *Gastrophryne*. Cat. Am. Amphib. Reptiles 134.1-134.2.—Range map and literature review. Bibliographies, Distribution. *Gastrophryne*.

Nelson, J., *see* 376

Netting, M. G., *see* 702

1112 **Newman, G. A.** 1970. Cowbird parasitism and nesting success of lark sparrows in southern Oklahoma. Wilson Bull. 82(3):304-309.—Snakes were suspected of nest predation. Natural History. *Coluber, Elaphe*.

1113 **Nice, M. M.** 1921. Our greatest food loss—shall we have cats or birds? Proc. Okla. Acad. Sci. 1:48.—Author discussed the economic value of toads. Natural History. *Bufo*.

1114 **Nice, M. M.** 1921. Some experiences with mourning doves in captivity. Proc. Okla. Acad. Sci. 1:57-65.—Mentions doves' reaction to hopping toads and frogs. Natural History. *Bufo, Rana*.

1115 **Nice, M. M.** 1927. The bird life on a forty acre tract in central Oklahoma. Proc. Okla. Acad. Sci. 7:25-93.—Frogs and snakes were mentioned. Natural History. *Rana*.

1116 **Oelrich, T. M.** 1957. The status of the Upper Pliocene turtle *Testudo turgida* Cope. J. Paleontol. 31(1):228-241.—Oklahoma fossils included. General Reference, Paleontology.

Oldham, R. S., *see* 969

1117 **Oliver, G. V., Jr.** 1970. Black ratsnake predation upon nesting barn and cliff swallows. Bull. Okla. Ornithol. Soc. 3(3):17-20.—Natural History. *Elaphe*.

Oliver, G. V., Jr., *see* 720

1118 **Oliver, J. A.** 1955. The natural history of North American amphibians and reptiles. D. Van Nostrand Company, Inc., New York. 359 pp.—General reference to Oklahoma. Natural History. All Amphibia, All Reptilia.

1119 **Olson, E. C.** 1941. The family Trematopsidae. J. Geol. 49(2):149-176.—Fossil amphibians from Oklahoma included. General Reference, Paleontology.

1120 **Olson, E. C.** 1954. Fauna of the Vale and Choza: 9, *Captorhinomorpha.* Fieldiana Geol. 10(19):211-218.—Fossil reptiles found in fissure fills at Richard's Spur, Oklahoma. General Reference, Paleontology.

1121 **Olson, E. C.** 1954. Notes on the stratographic and geographic ranges of certain genera of Permian vertebrates. J. Geol. 62(6):610-611.—Fossil amphibians and reptiles from Oklahoma included. General Reference, Paleontology.

1122 **Olson, E. C.** 1958. Parallelism in the evolution of the Permian reptilian faunas of the old and new worlds. Fieldiana Zool. 37(13):385-401.—General Reference, Paleontology.

1123 **Olson, E. C.** 1962. Late Permian terrestrial vertebrates, U.S.A. and U.S.S.R. Trans. Am. Philos. Soc. 52(2):1-224.—Amphibians and reptiles from Oklahoma are included. Distribution, General Reference, Paleontology.

1124 **Olson, E. C.** 1962. Permian vertebrates from Oklahoma and Texas. Part II. Osteology of *Captorhinkas* Olson. Okla. Geol. Surv. Circ. (59):49-68.—Fossil reptiles included, General Reference, Paleontology.

1125 **Olson, E. C.** 1965. New Permian vertebrates from the Chickasha Formation in Oklahoma. Okla. Geol. Surv. Circ. (70):1-70.—Fossil amphibians and reptiles were found. General Reference, Paleontology.

1126 **Olson, E. C.** 1967. Early Permian vertebrates of Oklahoma. Okla. Geol. Surv. Circ. (74):1-70.—Both amphibians and reptiles included. General Reference, Paleontology.

1127 **Olson, E. C.** 1968. The family Caseidae. Fieldiana Geol. 17(3):225-349.—Pelycosaurian reptiles from the Permian of Oklahoma. General Reference, Paleontology.

1128 **Olson, E. C.** 1970. New and little known genera and species of vertebrates from the Lower Permian of Oklahoma. Fieldiana Geol. 18(3):359-434.—Amphibians and reptiles from various fossil sites in Oklahoma. General Reference, Paleontology.

1129 **Olson, E. C.** 1971. A skeleton of *Lysorophus tricarinatus* (Amphibia: Lepospondyli) from the Hennessey Formation (Permian) of Oklahoma. J. Paleontol. 45(3):443-449.—General Reference, Paleontology.

1130 **Olson, E. C.** 1972. *Diplocaulus parvus* n. sp. (Amphibia: Nectridea) from the Chickasha Formation (Permian: Guadalupian) of Oklahoma. J. Paleontol. 46(5):656-659.—General Reference, Paleontology.

1131 **Olson, E. C.** 1972. *Fayella chickashaensis,* the dissorophoidea and the Permian terrestrial radiations. J. Paleontol. 46(1):104-114.—New amphibian from Chickasha Formation of Oklahoma. General Reference, Paleontology.

1132 **Olson, E. C.** 1972. Vertebrates from the Chickasaw Formation of Oklahoma. Soc. Vert. Paleontol. Ann. Meeting.—General Reference, Paleontology.

1133 **Olson, E. C.** 1979. *Seymouria grandis* n. sp. (Batrachosaura: Amphibia) from the Middle Clear Fork (Permian) of Oklahoma and Texas. J. Paleontol. 53(3):720-728.—General Reference, Paleontology.

1134 **Olson, E. C.** 1984. The taxonomic status and morphology of *Pleuristion brachyceolus* Case; referred to *Protocaptarhinus pricei* Clark and Carroll (Reptilia: Captorhinomorpha). J. Paleontol. 58(5):1282-1295.—The specimen was from Logan County, Oklahoma. General Reference, Paleontology.

1135 **Olson, E. C., and H. Barghusen.** 1962. Permian reptiles from Oklahoma and Texas Part I. Vertebrates from the Flowerpot Formation, Permian of Oklahoma. Okla. Geol. Surv. Circ. (59):1-48.—Fossil amphibians and reptiles described. General Reference, Paleontology.

1136 **Olson, E. C., and C. C. Branson.** 1965. *Zatrachys serratus* Cope (Amphibia: Labyrinthodontia) from McClain County, Oklahoma. Okla. Geol. Notes 25(4):91-97.—General Reference, Paleontology.

1137 **Olson, E. C., and R. L. Miller.** 1951. A mathematical model applied to study of evolution of species. Evolution 5(4):325-338.—Samples included fossil reptiles from Oklahoma. General Reference, Paleontology.

1138 **Olson, E. C., and P. P. Vaughn.** 1970. The changes of terrestrial vertebrates and climates during the Permian of North America. Forma Functio. 3:113-138.—Fossil amphibians and reptiles from Oklahoma included. Distribution, Ecology, General Reference, Paleontology.

1139 **Ortenburger, A. I.** 1923. The status of *Sonora semiannulata* and *Sonora episcopa.* Copeia 1923(120):79-81.—Morphology, Taxonomy. *Sonora.*

1140 **Ortenburger, A. I.** 1924. Life history notes - *Scaphiopus* - the spadefoot toad. Proc. Okla. Acad. Sci. 4:19-20.—Behavior, Natural History. *Scaphiopus.*

1141 **Ortenburger, A. I.** 1925. A preliminary list of the snakes of Oklahoma. Proc. Okla. Acad. Sci. 5:83-87.—Distribution. *Arizona, Carphophis, Cemophora, Coluber, Diadophis, Elaphe, Heterodon, Lampropeltis, Masticophis, Nerodia, Opheodrys, Pituophis, Regina, Sonora, Storeria, Tantilla, Thamnophis, Tropidoclonion, Virginia, Leptotyphlops, Agkistrodon, Crotalus, Sistrurus.*

1142 **Ortenburger, A. I.** 1926. Exploring Oklahoma's unknown. Outdoor Okla. 2(10):6-7, 30-31.—Historical, Popular. *Bufo, Acris, Rana, Chelydra, Ophisaurus, Crotaphytus, Phrynosoma, Sceloporus, Eumeces, Scincella, Coluber, Masticophis, Nerodia, Pituophis, Thamnophis, Crotalus.*

1143 **Ortenburger, A. I.** 1926. A key to the snakes of Oklahoma. Proc. Okla. Acad. Sci. 6:197-218.—Taxonomy. *Arizona, Cemophora, Coluber, Diadophis, Farancia, Heterodon, Lampropeltis, Masticophis, Nerodia, Rhinocheilus, Storeria, Tantilla, Thamnophis, Virginia, Leptotyphlops, Agkistrodon, Crotalus, Sistrurus.*

1144 **Ortenburger, A. I.** 1926. A report on the amphibians and reptiles of Oklahoma. Proc. Okla. Acad. Sci. 6:89-100.—Distribution. *Bufo, Acris, Pseudacris, Gastrophryne, Scaphiopus, Rana, Ambystoma, Desmognathus, Notophthalmus, Chelydra, Chrysemys, Deirochelys, Graptemys, Pseudemys, Trachemys, Terrapene, Kinosternon, Sternotherus, Trionyx, Ophisaurus, Anolis, Crotaphytus, Holbrookia, Phrynosoma, Sceloporus, Eumeces, Scincella, Cnemidophorus, Arizona, Carphophis, Cemophora, Coluber, Diadophis, Elaphe, Heterodon, Lampropeltis, Masticophis, Nerodia, Opheodrys, Pituophis, Regina, Sonora, Storeria, Tantilla, Thamnophis, Tropidoclonion, Virginia, Leptotyphlops, Agkistrodon, Crotalus, Sistrurus.*

1145 **Ortenburger, A. I.** 1926. Reptiles and amphibians collected in the Arbuckle Mts., Murray County, Oklahoma. Copeia 1926(156):145-146.—Distribution. *Bufo, Acris, Rana, Chelydra, Crotaphytus, Sceloporus, Cnemidophorus, Coluber, Diadophis, Elaphe, Lampropeltis, Masticophis, Nerodia, Opheodrys, Tantilla, Thamnophis.*

1146 **Ortenburger, A. I.** 1926. Reptiles and amphibians collected in the Wichita Mts., Comanche County, Oklahoma. Copeia 1926(155):137-138.—Distribution. *Bufo, Acris, Rana, Chelydra, Crotaphytus, Phrynosoma, Sceloporus, Eumeces, Coluber, Masticophis, Nerodia, Pituophis, Storeria, Thamnophis, Crotalus, Sistrurus.*

1147 **Ortenburger, A. I.** 1926. The University of Oklahoma Museum of Zoology expedition of 1925. Proc. Okla. Acad. Sci. 6:85-88.—Lists approximate numbers of amphibians and reptiles collected and placed in museum. General Reference, Historical.

1148 **Ortenburger, A. I.** 1927. A list of reptiles and amphibians from the Oklahoma Panhandle. Copeia 1927(163):46-48.—Distribution. *Bufo, Scaphiopus, Rana, Ambystoma, Chelydra, Trachemys, Terrapene, Kinosternon, Crotaphytus, Holbrookia, Phrynosoma, Sceloporus, Eumeces, Cnemidophorus, Coluber, Heterodon, Lampropeltis, Masticophis, Nerodia, Pituophis, Thamnophis, Crotalus.*

1149 **Ortenburger, A. I.** 1928. The whip snakes and racers: Genera *Masticophis* and *Coluber.* Mem. Univ. Mich. Mus. (1):1-247.—Distribution, Evolution, Morphology, Taxonomy. *Coluber, Masticophis.*

1150 **Ortenburger, A. I.** 1929. Reptiles and amphibians from northeastern Oklahoma. Copeia 1929(170):26-28.—Distribution. *Bufo, Acris, Hyla, Gastrophryne, Rana, Eurycea, Necturus, Notophthalmus, Graptemys, Terrapene, Trionyx, Crotaphytus, Phrynosoma, Sceloporus, Eumeces, Scincella, Cnemidophorus, Coluber, Diadophis, Masticophis, Nerodia, Opheodrys, Storeria, Thamnophis, Agkistrodon.*

1151 **Ortenburger, A. I.** 1929. Reptiles and amphibians from southeastern Oklahoma and southwestern Arkansas. Copeia 1929(170):8-12.—Distribution. *Bufo, Acris, Hyla, Pseudacris, Gastrophryne, Rana, Desmognathus, Eurycea, Plethodon, Terrapene, Trionyx, Crotaphytus, Sceloporus, Eumeces, Scincella, Cnemidophorus, Carphophis, Coluber, Elaphe, Heterodon, Lampropeltis, Masticophis, Nerodia, Opheodrys, Thamnophis, Agkistrodon, Crotalus.*

1152 **Ortenburger, A. I.** 1929. Snakes: harmless and poisonous. Sooner Mag. 2(3):90.—Lore, Popular. *Cemophora, Elaphe, Heterodon, Masticophis, Pituophis, Crotalus, Sistrurus.*

1153 **Ortenburger, A. I.** 1930. A key to the lizards and snakes of Oklahoma. Publ. Univ. Okla. Biol. Surv. 2(4):209-239.—Taxonomy. *Ophisaurus, Anolis, Crotaphytus, Holbrookia, Phrynosoma, Sceloporus, Eumeces, Scincella, Cnemidophorus, Arizona, Carphophis, Cemophora, Coluber, Diadophis, Elaphe, Farancia, Heterodon, Lampropeltis, Masticophis, Nerodia, Opheodrys, Pituophis, Regina, Rhinocheilus, Sonora, Storeria, Tantilla, Thamnophis, Tropidoclonion, Virginia, Leptotyphlops, Agkistrodon, Crotalus, Sistrurus.*

1154 **Ortenburger, A. I.** 1930. Reptiles and amphibians from Pawnee County, Oklahoma. Copeia 1930(173):94-95.—Distribution. *Bufo,*

Acris, Ophisaurus, Crotaphytus, Phrynosoma, Eumeces, Scincella, Cnemidophorus, Coluber, Elaphe, Lampropeltis, Masticophis, Opheodrys, Pituophis, Storeria, Tantilla, Thamnophis, Tropidoclonion.

1155 **Ortenburger, A. I.** 1930. Some animals of the Arbuckles. Southwest Wild Waters 2(1):30-31.—Popular. *Bufo, Acris, Rana, Chelydra, Terrapene, Crotaphytus, Sceloporus, Scincella, Cnemidophorus, Coluber, Diadophis, Elaphe, Heterodon, Lampropeltis, Opheodrys, Sonora, Storeria, Tantilla, Thamnophis, Agkistrodon, Crotalus.*

1156 **Ortenburger, A. I.** 1930. Some common snake stories. School Sci. Math. 30(4):420-428.—Popular. *Elaphe.*

1157 **Ortenburger, A. I.** 1952. Treatment of snake bites. J. Okla. State Med. Assoc. 45(6):1-5.—Techniques. *Agkistrodon, Crotalus, Sistrurus.*

1158 **Ortenburger, A. I., and R. D. Bird.** 1933. The ecology of the western Oklahoma Salt Plains. Publ. Univ. Okla. Biol. Surv. 5(3):49-64.—Distribution. *Bufo, Chelydra, Terrapene, Holbrookia, Sceloporus, Eumeces, Cnemidophorus, Arizona, Heterodon, Pituophis, Rhinocheilus, Sistrurus.*

1159 **Ortenburger, A. I., and B. Freeman.** 1930. Notes on some reptiles and amphibians from western Oklahoma. Publ. Univ. Okla. Biol. Surv. 2(4):175-188.—Distribution. *Bufo, Acris, Pseudacris, Scaphiopus, Rana, Ambystoma, Chelydra, Trachemys, Terrapene, Kinosternon, Trionyx, Oph isaurus, Crotaphytus, Holbrookia, Phrynosoma, Sceloporus, Eumeces, Scincella, Cnemidophorus, Coluber, Diadophis, Elaphe, Heterodon, Lampropeltis, Masticophis, Nerodia, Opheodrys, Pituophis, Rhinocheilus, Sonora, Storeria, Thamnophis, Leptotyphlops, Agkistrodon, Crotalus.*

Ortenburger, A. I., *see* 1085, 1481

1160 **Overdeer, D. C.** 1979. Effects of stream alteration on amphibian and reptile populations in selected habitats in southern Oklahoma. M.S. Thesis, Okla. State Univ., Stillwater. 19 pp.—Natural History. *Acris, Rana, Scincella, Nerodia.*

1161 **Overdeer, D. C.** 1983. Reptiles of the mesquite grasslands of southwest Oklahoma. D.Ed. Diss., Okla. State Univ., Stillwater. 34 pp.—Distribution, Natural History. *Chelydra, Chrysemys, Terrapene, Kinosternon, Trionyx, Crotaphytus, Holbrookia, Phrynosoma, Sceloporus, Uta, Eumeces, Scincella, Cnemidophorus, Arizona, Coluber, Diadophis, Elaphe, Heterodon, Hypsiglena, Lampropeltis, Masticophis, Nerodia, Pituophis, Regina, Rhinocheilus, Sonora, Tantilla, Thamnophis, Leptotyphlops, Crotalus, Sistrurus.*

1162 **Owings, D. H., and S. C. Owings.** 1979. Snake-directed behavior by black-tailed prairie dogs (*Cynomys ludovicianus*). Z. Tierpsychol. 49(1):35-54.—Study conducted at the Wichita Mountains Wildlife Refuge and at the Oklahoma City Zoo. Behavior. *Crotalus, Pituophis.*

1163 **Owings, D. H., and W. J. Loughry.** 1985. Variation in snake-elicited jump-yipping by black-tailed prairie dogs: Ontogeny and snake-specificity. Z. Tierpsychol. 70(3):177-200.—Study conducted at Wichita Mountains National Wildlife Refuge. Behavior. *Crotalus, Pituophis.*

Owings, S. C., *see* 1162

1164 **Pace, A. E.** 1974. Systematic and biological studies of the leopard frogs (*Rana pipiens* complex) of the United States. Misc. Publ. Mus. Zool. Univ. Mich. (148):1-140.—Behavior, Bibliographies, Distribution, Morphology, Natural History, Taxonomy. *Rana.*

Palmer, E. M., *see* 151, 975

1165 **Palmer, W. M.** 1978. *Sistrurus miliarius.* Cat. Am. Amphib. Reptiles 220.1-220.2.—Range map and literature review. Bibliographies, Distribution. *Sistrurus.*

1166 **Parker, J. W.** 1982. Opportunistic feeding by an ornate box turtle under a nest of a Mississippi kite. Southwest. Nat. 27(3):365.— Natural History. *Terrapene.*

1167 **Parmley, D. C., and J. D. Tyler.** 1978. Gray treefrog in southwestern Oklahoma. Southwest. Nat. 23(1):157-158.—Distribution. *Hyla.*

1168 **Parmley, D. C., and J. D. Tyler.** 1978. Herpetofaunal prey at a burrowing owl hole. Bull. Okla. Herp. Soc. 3(1):13.—Natural History. *Rana, Ambystoma, Phrynosoma, Leptotyphlops.*

Parmley, D. C., *see* 1446

1169 **Parrish, H. M., and C. A. Carr.** 1967. Bites of copperheads (*Ancistrodon contortrix*) in the United States. J. Am. Med. Assoc. 201(12): 107-112.—Health. *Agkistrodon.*

1170 **Patterson, P.** 1945. Snake! Okla. Game Fish News 1(4-5):10-11.— Popular. *Agkistrodon, Crotalus, Sistrurus.*

1171 **Patton, L. J.** 1978. Checkered garter snakes found. Bull. Okla. Herp. Soc. 3(3):53.—Reprinted from Caldwell Elementary School Science Newsletter. Distribution. *Thamnophis.*

1172 **Paulissen, M. A.** 1985. Social behavior and thermoregulation in male collared lizards. Bull. Okla. Herp. Soc. 10(3-4):17-20.—Behavior, Physiology. *Crotaphytus.*

1173 **Paulissen, M. A.** 1986. Home range of adult and juvenile six-lined racerunners. Bull. Okla. Herp. Soc. 11(1-4):1-5.—Behavior, Ecology. *Cnemidophorus.*

1174 **Paulissen, M. A.** 1986. Ontogenetic changes in the ecology of the six-lined racerunner, *Cnemidophorus sexlineatus* (Sauria: Teiidae). Ph.D. Diss., Univ. Okla., Norman. 129 pp.—Behavior, Ecology. *Cnemidophorus.*

1175 **Paulissen, M. A.** 1986. A technique for marking teiid lizards in the field. Herpetol. Rev. 17(1):16-17.—Techniques. *Cnemidophorus.*

1176 **Paulissen, M. A.** 1987. Diet of adult and juvenile six-lined racerunners, *Cnemidophorus sexlineatus* (Sauria: Teiidae). Southwest. Nat. 32(3):395-397.—Ecology. *Cnemidophorus.*

1177 **Paulissen, M. A.** 1987. Optimal foraging and intraspecific diet differences in the lizard *Cnemidophorus sexlineatus.* Oecologia 71(3): 439-446—Ecology. *Cnemidophorus.*

1178 **Paulissen, M. A., and B. C. Harvey.** 1985. *Cnemidophorus sexlineatus viridis* (prairie-lined racerunner). Arboreality. Herpetol. Rev. 16(1): 27.—Natural History. *Cnemidophorus.*

Paulissen, M. A., *see* 1459

1179 **Paulson, B. K.** 1986. The origin of the stimulus from muscle spasms and blood changes which occur at the critical thermal maximum in anurans. Ph.D. Diss., Univ. Okla., Norman. 47 pp.—Physiology. *Bufo.*

1180 **Paulson, B. K., and V. H. Hutchison.** 1987. Blood changes in *Bufo cognatus* following acute heat stress. Comp. Biochem. Physiol. 87(A):461-466.—Physiology. *Bufo.*

1181 **Paulson, B. K., and V. H. Hutchison.** 1987. Origin of the stimulus for muscular spasms at the critical thermal maximum in anurans. Copeia 1987(3):810-813.—Physiology. *Bufo.*

1182 **Peabody, F. E.** 1951. The origin of the astragulus of reptiles. Evolution 5(4):339-344.—Specimens included from Lower Permian of Fort Sill. General Reference, Morphology, Paleontology.

1183 **Peabody, F. E.** 1961. Annual growth zones in living and fossil vertebrates. J. Morphol. 108(1):11-62.—Microsaur amphibians of the

Lower Permian fissure deposits of Fort Sill were used. General Reference, Paleontology.

Peabody, F. E., *see* 722

1184 **Penman, J. T.** 1974. Faunal analysis of the Hodge Site, Custer County, Oklahoma. Pages 69-72 *in* Okla. Contract Archeol. Surv. 1: 1-163.—Archeology. *Terrapene, Kinosternon, Trionyx.*

1185 **Penman, J. T.** 1975. Faunal interpretations in Fort Washita from past to present: An archaeological report. Pages 270-273 *in* Okla. Hist. Soc. Ser. Anthropol. (1):1-287.—Archeology. *Terrapene.*

1186 **Penman, J. T.** 1979. Faunal remains from the Pate Site, Oklahoma. Bull. Okla. Anthropol. Soc. 28:147-150.—Archeology. *Terrapene, Trionyx.*

1187 **Perino, G., and W. J. Bennett, Jr.** 1978. Archeological investigations at the Mahaffey Site Ch-1, Hugo Reservoir, Choctaw County, Oklahoma. Museum of the Red River publication, Idabel. 110 pp.—Turtle shells were found. Archeology. *Graptemys, Trachemys, Terrapene.*

1188 **Perkins, C. B.** 1940. A key to the snakes of the United States. Bull. Zool. Soc. San Diego (16):1-63.—Distribution, Taxonomy. All Serpentes.

1189 **Peterka, H. E.** 1940. The anatomy of an anomalous thoracic region in *Rana catesbeiana.* Proc. Okla. Acad. Sci. 21:47-48.—Morphology. *Rana.*

1190 **Peters, J. A.** 1952. Catalogue of type specimens in the herpetological collections of the University of Michigan Museum of Zoology. Occas. Pap. Mus. Zool. Univ. Mich. (539):1-55.—Taxonomy. *Eurycea, Typhlotriton.*

1191 **Peyer, B.** 1968. Comparative odontology. Univ. Chicago Press. 336 pp.—Oklahoma reptiles included. General Reference, Paleontology.

1192 **Phillips, P., and J. A. Brown.** 1978. Pre-Columbian shell engravings from the Craig Mound at Spiro, Oklahoma. 2 Parts. Harvard Univ. Press, Cambridge. 1097 pp.—Archeology. *Crotalus.*

1193 **Pickering, D.** 1982. Notes on reproduction of the western worm snake. Bull. Okla. Herp. Soc. 7(3-4):86-88.—Natural History. *Carphophis.*

1194 **Pickering, D.** 1984. Box turtle's cracked shell repaired at the Tulsa Zoo. Bull. Okla. Herp. Soc. 9(3):28-29.—Techniques. *Terrapene.*

Pierce, J. R., *see* 403

1195 **Pietruszka, R. D.** 1981. Use of scutellation for distinguishing sexes in bisexual species of *Cnemidophorus.* Herpetologica 37(4):244-249.—Morphology. *Cnemidophorus.*

Pigg, J., *see* 94, 154, 664, 674

1196 **Pisani, G. R., J. T. Collins, and S. R. Edwards.** 1972. A re-evaluation of the subspecies of *Crotalus horridus.* Trans. Kans. Acad. Sci. 75(3): 255-263.—Distribution, Taxonomy. *Crotalus.*

1197 **Platt, D. R.** 1969. Natural history of the hognose snakes *Heterodon platyrhinos* and *Heterodon nasicus.* Univ. Kans. Publ. Mus. Nat. Hist. 18(4):253-420.—Bibliographies, Distribution, Natural History. *Heterodon.*

1198 **Platt, D. R.** 1983. *Heterodon.* Cat. Am. Amphib. Reptiles 315.1-315.2.—Range map and literature review. Bibliographies, Distribution. *Heterodon.*

Plsek, R. W., *see* 1470

1199 **Pollard, L.** 1985. Oklahoma's poisonous snakes. Outdoor Okla. 41(2):14-21.—Natural History, Popular. *Agkistrodon, Crotalus, Sistrurus.*

1200 **Pope, C. H.** 1946. Turtles of the United States and Canada. Alfred A. Knopf, New York. 337 pp.—Distribution. *Chelydra, Macroclemys, Deirochelys, Graptemys, Terrapene, Kinosternon, Sternotherus, Trionyx.*

1201 **Pope, C. H., and S. H. Pope.** 1951. A study of the salamander *Plethodon ouachitae* and a description of an allied form. Bull. Chic. Acad. Sci. 9(8):129-152.—Distribution, Ecology, Natural History. *Plethodon.*

Pope, S. H., *see* 1201

Pratt, W. L., Jr., *see* 1202

1202 **Preston, J. R., and W. L. Pratt, Jr.** 1962. An eastward range extension of *Uta stansburiana stejnegeri.* Herpetologica 18(1):53.—Distribution. *Uta.*

1203 **Preston, R. E.** 1980. Late Pleistocene cold-blooded vertebrate faunas from the mid-continental United States. Pap. Paleontol. Mus. Paleontol. Univ. Mich. (19):1-51.—General Reference, Paleontology.

1204 **Preston, R. E., and C. J. McCoy, Jr.** 1971. The status of *Emys twentii* Taylor (Reptilia: Testudinidae) based on new fossil records from Kansas and Oklahoma. J. Herpetol. 5(1-2):23-30.—General Reference, Paleontology.

1205 **Preston, W. B.** 1970. The comparative ecology of two water snakes, *Natrix rhombifera* and *Natrix erythrogaster,* in Oklahoma. Ph.D. Diss., Univ. Okla., Norman. 79 pp.—Behavior, Ecology. *Nerodia.*

Preston, W. B., *see* 844

1206 **Prewitt, D.** 1978. Animals collected in Grant County, Oklahoma. Bull. Okla. Herp. Soc. 3(3):54.—Reprinted from Caldwell Elementary School Science Newsletter. Distribution. *Trachemys, Lampropeltis, Nerodia.*

1207 **Price, A. H.** 1986. *Cnemidophorus tesselatus.* Cat. Am. Amphib. Reptiles 398.1-398.2.—Range map and literature review. Bibliographies, Distribution. *Cnemidophorus.*

Price, I. W., *see* 1244

1208 **Price, L. I.** 1935. Notes on the braincase of *Captorhinus.* Boston Soc. Nat. Hist. Proc. 40:377-385.—Oklahoma reptiles were included. General Reference, Paleontology.

1209 **Price, L. I.** 1940. Autotomy of the tail in Permian reptiles. Copeia 1940(2):119-120.—Caudal vertebrae of *Captorhinus* were from Oklahoma. General Reference, Paleontology.

Price, L. I., *see* 722, 1378

1210 **Proctor, V.** 1958. The growth of *Basicladia* on turtles. Ecology 39(4):634-645.—Article discusses a commensal alga that grows on turtles. Natural History. *Chelydra, Graptemys, Trachemys, Kinosternon.*

Puckette, W. L., *see* 156

1211 **Purdue, J. R.** 1970. A comparative study of the body movements of displaying males of the lizard genus *Sceloporus* (Iguanidae). M.S. Thesis, Univ. Okla., Norman. 57 pp.—Behavior. *Sceloporus.*

1212 **Purdue, J. R.** 1972. Herpetol. Rev. 4(2):27.—Photograph of a western spadefoot toad from Oklahoma. Behavior. *Scaphiopus.*

1213 **Purdue, J. R.** 1973. HISS News J. 1(4):123.—Photograph of a yellow mud turtle. Popular. *Kinosternon.*

1214 **Purdue, J. R., and C. C. Carpenter.** 1970. The phylogenetic implications of body movements in displaying sceloporine lizards. Herpetol. Rev. 2(4):7.—An Abstract. Behavior. *Sceloporus.*

1215 **Purdue, J. R., and C. C. Carpenter.** 1972. A comparative study of the body movements of displaying males of the lizard genus *Sceloporus* (Iguanidae). Behaviour 41(1-2):68-81.—Behavior, Techniques. *Sceloporus.*

1216 **Purdue, J. R., and C. C. Carpenter.** 1972. A comparative study of display motion in the iguanid genera *Sceloporus, Uta,* and *Urosaurus.* Herpetologica 28(2):137-140.—Behavior, Techniques. *Sceloporus.*

1217 **Quinn, H. R.** 1979. Sexual dimorphism in the tail pattern of Oklahoma snakes. Tex. J. Sci. 31(2):157-160.—Morphology. *Arizona, Cemophora, Diadophis, Elaphe, Farancia, Heterodon, Lampropeltis, Nerodia, Pituophis, Rhinocheilus, Sonora, Tropidoclonion, Agkistrodon, Crotalus, Sistrurus.*

1218 **Rachuk, N. A.** 1978. Genetic variation in *Sceloporus undulatus:* effects of gene flow, isolation, and selection. M.S. Thesis, Okla. State Univ., Stillwater. 103 pp.—Evolution, Genetics. *Sceloporus.*

1219 **Ralin, D. B.** 1968. Ecological and reproductive differentiation in cryptic species of the *Hyla versicolor* complex (Hylidae). Southwest. Nat. 13(3):283-300.—Behavior, Ecology, Natural History. *Hyla.*

1220 **Ralin, D. B.** 1977. Evolutionary aspects of mating call variation in a diploid-tetraploid species complex of treefrogs (Anura). Evolution 31(4):721-736.—Evolution, Natural History, Taxonomy. *Hyla.*

1221 **Ramsey, L. W.** 1953. The lined snake, *Tropidoclonion lineatum* (Hallowell). Herpetologica 9(1):7-24.—Bibliographies, Distribution, Ecology, Morphology, Natural History, Taxonomy. *Tropidoclonion.*

1222 **Rankin, E. J.** 1976. Rattlesnake round-up. Bull. Okla. Herp. Soc. 1(3):38.—An editorial on the cruelty of rattlesnake hunts. Management. *Crotalus.*

1223 **Ray, G. E.** 1941. Big for his day. Nat. Hist. 48(1):36-39.—A popular article on Oklahoma dinosaurs. General Reference, Paleontology, Popular.

1224 **Ray, G. E.** 1958. Spying for college credit. Ford Times 50(6):27-29.—An article about life at the Oklahoma Biological Station that included catching snakes. General Reference, Popular.

1225 **Ray, G. E.** 1960. Life with lizards. Sci. Digest 47(1):20-23.—Behavior, Popular. *Crotaphytus, Phrynosoma, Cnemidophorus.*

1226 **Ray, M. A.** 1965. A report of excavations at the Pohly Site, MY-54, in northeastern Oklahoma. Bull. Okla. Anthropol. Soc. 13:1-68.—"Turtle shells" found in the site. Archeology, General Reference.

1227 **Raymond, L. R., and L. M. Hardy.** 1983. Taxonomic status of the corn snake, *Elaphe guttata* (Linnaeus) (Colubridae), in Louisiana and eastern Texas. Southwest. Nat. 28(1):105-107.—Article includes a range map with Oklahoma locations. Distribution. *Elaphe.*

1228 **Reese, R. W., and H. M. Smith.** 1951. Pattern neoteny in the salamander *Eurycea lucifuga* Rafinesque. Copeia 1951(3):243-244.—Adults show a color pattern characteristic of the larval stage. Morphology. *Eurycea.*

1229 **Reeve, W. L.** 1952. Taxonomy and distribution of the horned lizards genus *Phrynosoma.* Univ. Kans. Sci. Bull. 34(14):817-960.—Distribution, Taxonomy. *Phrynosoma.*

1230 **Reisz, R. R.** 1980. A protorothyridid captorhinomorph reptile from the Lower Permian of Oklahoma. Life Sci. Contrib. R. Ont. Mus. (121): 1-16.—General Reference, Paleontology.

Reisz, R. R., *see* 751

Reno, H. W., *see* 832, 881

1231 **Reynolds, S. L., and M. E. Seidel.** 1982. *Sternotherus odoratus.* Cat. Am. Amphib. Reptiles 287.1-287.4.—Range map and literature review. Bibliographies, Distribution. *Sternotherus.*

Richards, A., *see* 386, 629

1232 **Richards, M. K.** 1971. The Lee Site, a late prehistoric manifestation in Garvin County, Oklahoma. Bull. Okla. Anthropol. Soc. 20:1-82.—Archeology. *Graptemys, Trachemys, Trionyx.*

Richter, K. E., *see* 1414

1233 **Rider, E.** 1979. Winter activity in the red-eared turtle. Bull. Okla. Herp. Soc. 4(2-3):39.—Natural History. *Trachemys.*

Rigney, C. C., *see* 1075

1234 **Riley, J., and J. T. Self.** 1979. On the systematics of the pentastomid genus *Porocephalus* (Humboldt, 1811) with description of two new species. Syst. Parasitol. 1(1):25-42.—Parasitology. *Crotalus.*

1235 **Rinker, G. C., and C. W. Hibbard.** 1952. A new beaver and associated vertebrates from the Pleistocene of Oklahoma. J. Mammal. 33(1):98-101.—An emydid turtle was included. General Reference, Paleontology.

Rogers, J. S., *see* 673

1236 **Rodgers, L. O.** 1941. *Diplorchis scaphiopi,* a new polystomatid monogenean fluke from the spadefoot toad. M.S. Thesis, Okla. A.M. Coll., Stillwater.—Only available copy is a bound reprint from the Journal of Parasitology. Parasitology. *Scaphiopus.*

1237 **Rodgers, L. O.** 1941. *Diplorchis scaphiopi,* a new polystomatid monogenean fluke from the spadefoot toad. J. Parasitol. 27(2):153-157.—Parasitology. *Scaphiopus.*

1238 **Rodgers, L. O., and R. E. Kuntz.** 1940. A new polystomatid monogenean fluke from a spadefoot. Wasmann Club Coll. 4(2):37-40.—Parasitology. *Scaphiopus.*

1239 **Rodrick, G. E.** 1971. An ontogenetic study of lactate dehydrogenase in *Porocephalus crotali* (Pentastomidae). Ph.D. Diss., Univ. Okla., Norman. 68 pp.—Parasitology. *Crotalus.*

1240 **Rogers, J. S.** 1972. Discriminate function analysis of morphological relationships within the *Bufo cognatus* species group. Copeia 1972(2): 381-383.—Morphology. *Bufo.*

1241 **Rogers, T. E.** 1941. A study of the nucleolus of *Scaphiopus bombifrons* Cope. M.S. Thesis, Univ. Okla., Norman. 27 pp.—Morphology. *Scaphiopus.*

1242 **Romer, A. S.** 1956. The osteology of reptiles. Univ. Chicago Press. 772 pp.—Oklahoma reptiles were included. General Reference, Paleontology.

1243 **Romer, A. S.** 1966. Vertebrate paleontology. 3rd ed. Univ. Chicago Press. 468 pp.—Oklahoma amphibians and reptiles included. General Reference, Paleontology.

1244 **Romer, A. S., and I. W. Price.** 1940. Review of the Pelycosauria. Geol. Soc. Am. Spec. Pap. (28):1-538.—Oklahoma reptiles included. General Reference, Paleontology.

Romer, A. S., *see* 1378

1245 **Rossman, D. A.** 1963. The colubrid snake genus *Thamnophis:* A revision of the *sauritus* group. Bull. Fla. State Mus. Biol. Sci. 7(3):99-178. —Distribution, Taxonomy. *Thamnophis.*

1246 **Rossman, D. A.** 1970. *Thamnophis proximus.* Cat. Am. Amphib. Reptiles 98.1-98.3.—Range map and literature review. Bibliographies, Distribution. *Thamnophis.*

1247 **Rossman, D. A., and R. L. Erwin.** 1980. Geographic variation in the snake *Storeria occipitomaculata* (Storer) (Serpentes: Colubridae) in southeastern United States. Brimleyana 4:95-102.—Distribution, Taxonomy. *Storeria.*

1248 **Rostker, M. A.** 1983. An experimental study of collared lizards: Effects of habitat and male quality on fitness. Ph.D. Diss., Okla. State Univ., Stillwater. 62 pp.—Behavior, Evolution. *Crotaphytus.*

1249 **Rudolph, D. C.** 1980. Competition among five species of cave associated salamanders (family Plethodontidae). Ph.D. Diss. Tex. Tech. Univ., Lubbock. 74 pp.—Behavior, Ecology, Natural History. *Eurycea, Typhlotriton.*

Rush, T., *see* 686

1250 **Russell, D. T.** 1958. Guide to Robbers Cave State Park and Camp Tom Hall, Latimer County, Oklahoma. Pages 11-12 *in* Okla. Geol. Surv. Guidebook 7.—Natural History. *Trachemys, Pituophis, Agkistrodon, Crotalus.*

Rutherford, D. A., *see* 518

1251 **Ruthven, A. G.** 1908. Variations and genetic relationships of the garter-snakes. Bull. U.S. Natl. Mus. (61):1-201.—Distribution, Taxonomy. *Thamnophis.*

1252 **Salthe, S. N.** 1973. Amphiumidae. Cat. Am. Amphib. Reptiles 147.1-147.4—Range map and literature review. Bibliographies, Distribution. *Amphiuma.*

1253 **Salthe, S. N.** 1973. *Amphiuma tridactylum.* Cat. Am. Amphib. Reptiles 149.1-149.3.—Range map and literature review. Bibliographies, Distribution. *Amphiuma.*

1254 **Sanders, O.** 1953. A new species of toad, with a discussion of morphology of the bufonid skull. Herpetologica 9(1):25-47.—Bibliographies, Distribution, Morphology, Taxonomy. *Bufo.*

1255 **Sanders, O., and J. C. Cross.** 1963. Relationships between certain North American toads as shown by cytological study. Herpetologica 19(4):248-255.—Genetics, Morphology. *Bufo.*

1256 **Sanders, O., and H. M. Smith.** 1951. Geographic variation in toads of the *debilis* group of *Bufo*. Field Lab. 19(4):141-160.—Distribution, Taxonomy. *Bufo.*

Sanders, O., *see* 377

1257 **Sarjeant, W. A. S.** 1976. Track of a small amphibian from the Pennsylvanian of Oklahoma. Tex. J. Sci. 27(1):107-112.—General Reference, Paleontology.

1258 **Sassaman, J. F.** 1974. Heart rate of lizards of the families Iguanidae and Agamidae as an indicator of behavioral response to social stimuli. Ph.D. Diss., Univ. Okla., Norman. 86 pp.—Behavior. *Crotaphytus.*

1259 **Satterwhite, C. L.** 1959. Okla. Wildl. 15(7-8):17.—Photograph of diamondback rattlesnake killed for sport. Incorrectly printed as issue 4. Popular. *Crotalus.*

1260 **Sattler, P. W.** 1980. Genetic relationship among selected species of North American *Scaphiopus.* Copeia 1980(4):605-610.—Genetics. *Scaphiopus.*

1261 **Savage, D. E.** 1941. Two new Middle Pliocene carnivores from Oklahoma with notes on the Optima Fauna. Am. Midl. Nat. 25(3):692-710.—Paleontology. *Alligator.*

1262 **Savage, J. M.** 1954. A revision of the toads of the *Bufo debilis* complex. Tex. J. Sci. 6(1):83-112.—Bibliographies, Distribution, Evolution, Taxonomy. *Bufo.*

1263 **Schaaf, R. T., and P. W. Smith** 1970. Geographic variation in the pickerel frog. Herpetologica 26(2):240-254.—Distribution, Morphology. *Rana.*

1264 **Schaaf, R. T., and P. W. Smith.** 1971. *Rana palustris.* Cat. Am. Amphib. Reptiles 117.1-117.3.—Range map and literature review. Bibliographies, Distribution. *Rana.*

1265 **Schatti, B., and L. D. Wilson.** 1986. *Coluber.* Cat. Am. Amphib. Reptiles 399.1-399.4—Range map and literature review. Bibliographies, Distribution. *Coluber.*

Schemnitz, S. D., *see* 613

1266 **Schlemeyer, B.** 1957. Jackson County rattlers. Okla. Game Fish News 13(5):14.—Photograph of two diamondback rattlesnakes. Popular. *Crotalus.*

Schmerchel, N. L., *see* 653

1267 **Schmidt, F. L.** 1940. *Entonyssus vitzthumi* (Acarina), a new ophidian lung mite. J. Parasitol. 26(4):309-313.—Thirty-two mites were in a bull snake collected in Stillwater. Parasitology. *Pituophis.*

1268 **Schmidt, F. L.** 1946. *Entonyssus vitzthumi* (Acarina) a new ophidian lung mite. M.S. Thesis, Okla. A.M. Coll., Stillwater.—Only available copy is a bound reprint from the Journal of Parasitology. Parasitology. *Pituophis.*

1269 **Schmidt, K. P.** 1919. Rediscovery of *Amphiardis inornatus* (Garman), with notes on other specimens from Oklahoma. Copeia 1919(73):71-73.—Specimens were collected by a high school student. Distribution. *Bufo, Chelydra, Terrapene, Crotaphytus, Sceloporus, Elaphe, Heterodon, Lampropeltis, Virginia, Agkistrodon.*

1270 **Schmidt, K. P.** 1953. A checklist of North American amphibians and reptiles. 6th ed. Chicago Press. 280 pp.—Distribution, Taxonomy. *Bufo, Pseudacris, Gastrophryne, Scaphiopus, Rana, Ambystoma, Desmognathus, Eurycea, Plethodon, Typhlotriton, Necturus, Notophthalmus, Deirochelys, Graptemys, Terrapene, Kinosternon, Sternotherus, Holbrookia, Phrynosoma, Sceloporus, Eumeces, Scincella, Cnemidophorus, Arizona, Carphophis, Cemophora, Diadophis, Elaphe, Farancia, Hypsiglena, Lampropeltis, Masticophis, Nerodia, Opheodrys, Regina, Rhinocheilus, Sonora, Storeria, Tantilla, Thamnophis, Tropidoclonion, Virginia, Leptotyphlops, Agkistrodon, Crotalus, Sistrurus, Alligator.*

1271 **Schmidt, K. P., and D. D. Davis.** 1941. Field book of snakes of the United States and Canada. C. P. Putnam's Sons, New York. 305 pp. —Distribution. *Lampropeltis, Pituophis, Thamnophis, Crotalus.*

1272 **Schnell, G. D., F. L. Johnson, and J. L. Gentry, Jr.** 1979. Flora and fauna of Oklahoma abandoned mine lands. Rep. to Okla. Cons. Comm., Oklahoma City, from Okla. Biol. Surv. 135 pp.—Includes section on reptiles and amphibians of eastern Oklahoma and lists species by habitat. Distribution, Ecology. All Anura, *Ambystoma, Desmognathus, Eurycea, Plethodon, Typhlotriton, Necturus, Notophthalmus,* All Testudines, *Ophisaurus, Anolis, Crotaphytus, Holbrookia, Phrynosoma, Sceloporus, Eumeces, Scincella, Cnemidophorus, Carphophis, Coluber, Diadophis, Elaphe, Heterodon, Lampropeltis, Masticophis, Nerodia, Opheodrys, Pituophis, Regina, Sonora, Storeria, Tantilla, Thamnophis, Tropidoclonion, Virginia, Leptotyphlops, Agkistrodon, Crotalus, Sistrurus.*

Schnell, G. D., *see* 1288

1273 **Schultz, A. F.** 1925. Ecological succession in the Arbuckle region. Proc. Okla. Acad. Sci. 5:88-90.—Author claimed to have found pickerel frogs. Distribution. *Rana.*

1274 **Schultze, H. P., and J. Chorn.** 1983. A labyrinthodont palatine from the Permian of Fort Sill, Oklahoma, reinterpreted as a vomer. J. Paleontol. 57(5):1050-1052.—Included Oklahoma amphibians. General Reference, Paleontology.

1275 **Schwartz, A.** 1956. Geographic variation in the chicken turtle. Fieldiana Zool. 34(41):461-503.—Distribution, Morphology, Taxonomy. *Deirochelys.*

Schwartz, A., *see* 1534

1276 **Scott, G. W., and C. C. Carpenter.** 1955. Body temperatures of *Bufo w. woodhousei.* Proc. Okla. Acad. Sci. 36:84-85.—Ecology, Natural History, Physiology. *Bufo.*

1277 **Scott, N. J., Jr.** 1981. Status of the amphibians and reptiles of Arizona, New Mexico, Oklahoma and Texas. *In* Endangered species, region 2. 41 pp.—Bibliographies, Management. *Rana, Eurycea, Plethodon, Typhlotriton.*

1278 **Secor, S. M.** 1982. New additional records for Oklahoma reptiles. Bull. Okla. Herp. Soc. 7(3-4):81-82.—Distribution. *Chelydra, Crotaphytus, Sceloporus, Eumeces, Cnemidophorus, Cemophora, Elaphe, Lampropeltis, Pituophis, Sonora, Tantilla, Virginia, Agkistrodon, Crotalus.*

1279 **Secor, S. M.** 1983. *Lampropeltis getulus holbrooki* (speckled kingsnake). Reproduction. Herpetol. Rev. 14(1):20.—Natural History. *Lampropeltis.*

1280 **Secor, S. M.** 1984. Live herp exhibit at this year's joint ASIH-HL-SSAR meeting, Norman, Oklahoma. Bull. Okla. Herp. Soc. 9(3):22-25.—This is a list of the Oklahoma amphibians and reptiles displayed. General Reference.

1281 **Secor, S. M.** 1984. New county records for Oklahoma reptiles. Bull. Okla. Herp. Soc. 9(4):33-35.—Distribution. *Chrysemys, Deirochelys, Graptemys, Trionyx, Ophisaurus, Holbrookia, Eumeces, Carphophis, Elaphe, Lampropeltis, Opheodrys, Storeria, Tantilla, Tropidoclonion, Virginia.*

1282 **Secor, S. M.** 1985. Courtship and mating of the speckled kingsnake, *Lampropeltis getulus holbrooki.* M.S. Thesis, Univ. Okla., Norman. 55 pp.—Behavior, Techniques. *Lampropeltis.*

1283 **Secor, S. M.** 1986. The lesser siren, *Siren intermedia*, in Oklahoma. Bull. Okla. Herp. Soc. 11(1-4):25-31.—Behavior, Bibliographies, Distribution, Ecology, Morphology. *Siren.*

1284 **Secor, S. M.** 1986. Recent capture of the western lesser siren, *Siren intermedia nettingi*, in Oklahoma. Bull. Okla. Herp. Soc. 11(1-4):22-24.—Distribution, Ecology, Natural History. *Acris, Hyla, Rana, Ambystoma, Siren, Kinosternon, Sternotherus, Anolis, Eumeces, Scincella, Opheodrys, Agkistrodon.*

1285 **Secor, S. M.** 1986. Two new county records for reptiles in Oklahoma. Bull. Okla. Herp. Soc. 11(1-4):39.—Distribution. *Terrapene, Sceloporus.*

1286 **Secor, S. M.** 1987. Courtship and mating behavior of the speckled kingsnake, *Lampropeltis getulus holbrooki.* Herpetologica 43(1):15-28.—Behavior. *Lampropeltis.*

1287 **Secor, S. M., and C. C. Carpenter.** 1984. Distribution maps of Oklahoma reptiles. Okla. Herp. Soc. Spec. Publ. (3):1-57.—Distribution. All Testudines, All Sauria, All Serpentes.

1288 **Secor, S. M., and G. D. Schnell.** 1985. Evaluation of wetlands in Oklahoma's gulf coastal plain. Final Rep. to Dept. Wildl. Cons. Nongame Prog., Oklahoma City, Oklahoma. 110 pp.—Distribution. *Bufo, Hyla, Pseudacris, Gastrophryne, Rana, Ambystoma, Amphiuma, Desmognathus, Eurycea, Plethodon, Siren, Notophthalmus, Macroclemys, Chrysemys, Deirochelys, Graptemys, Pseudemys, Trachemys, Terrapene, Kinosternon, Sternotherus, Trionyx, Anolis, Sceloporus, Eumeces, Scincella, Cnemidophorus, Coluber, Diadophis, Elaphe, Heterodon, Lampropeltis, Masticophis, Nerodia, Opheodrys, Regina, Storeria, Tantilla, Thamnophis, Virginia, Agkistrodon, Crotalus, Alligator.*

Secor, S. M., *see* 157, 888

1289 **Seidel, M. E.** 1978. *Kinosternon flavescens.* Cat. Am. Amphib. Reptiles 216.1-216.4.—Range map and literature review. Bibliographies, Distribution. *Kinosternon.*

Seidel, M. E., *see* 1231

1290 **Seldon, D.** 1981. Scaly menagerie comes as jolt to man's visitors. Bull. Okla. Herp. Soc. 6(3):69.—General Reference, Popular.

1291 **Self, J. T.** 1938. Note on the food habits of *Chelydra serpentina.* Copeia 1938(4):200.—Natural History. *Rana, Chelydra.*

1292 **Self, J. T.** 1945. A new trematode, *Neorenifer crotali*, from the rattlesnake. J. Parasitol. 31(3):210-211.—Parasitology. *Crotalus.*

1293 **Self, J. T.** 1951. The auxilliary hooks in *Porocephalus crotali* Humboldt. Trans. Am. Microsc. Soc. 70(3):255-256.—Parasitology. *Crotalus.*

1294 **Self, J. T.** 1969. Biological relationships of the Pentastomida: A bibliography on the Pentastomida. Exp. Parasitol. 24(1):63-119.—Parasitology. *Crotalus.*

1295 **Self, J. T., and R. E. Kuntz.** 1967. Host-parasite relations in some Pentastomida. J. Parasitol. 53(1):202-206.—An Oklahoma diamondback rattlesnake was the host. Parasitology. *Crotalus.*

1296 **Self, J. T., and F. B. McMurry.** 1949. *Porocephalus crotali* Humboldt (Pentastomida) in Oklahoma. J. Parasitol. 34(1):21-23.—Parasitology. *Crotalus.*

Self, J. T., *see* 539, 548, 892, 1234, 1414

Sellards, E. H., *see* 495

1297 **Seltin, R. J.** 1950. A review of the family Captorhinidae. Fieldiana Geol. 10(34):461-500.—Fossil reptiles taken in Oklahoma. General Reference, Paleontology.

1298 **Sever, D. M.** 1980. Cloacal anatomy of male brook salamanders *(Eurycea).* Herpetologica 36(1):51-60.—Specimens were collected in Adair County. Morphology. *Eurycea.*

1299 **Shannon, F. A.** 1949. A western subspecies of *Bufo woodhousii* hitherto erroneously associated with *Bufo compactilis.* Bull. Chic. Acad. Sci. 8(15):301-312.—Oklahoma toads were used in the study. Taxonomy. *Bufo.*

Shannon, S. L., *see* 158, 159

1300 **Shelton, J. E.** 1976. Recognition of *Chrysemys concinna hieroglyphica* characteristics in Payne County, Oklahoma. Bull. Okla. Herp. Soc. 1(4):58-59.—Morphology. *Pseudemys.*

Sherrod, S. K., *see* 965

Shetlar, M. R., *see* 5

1301 **Shoop, C. R.** 1964. *Ambystoma talpoideum.* Cat. Am. Amphib. Reptiles 8.1-8.2.—Range map and literature review. Bibliographies, Distribution. *Ambystoma.*

Short, J., *see* 160

1302 **Sieg, E.** 1977. Red-tailed hawks and snake. Bull. Okla. Ornithol. Soc. 11(2):14.—Hawks were observed carrying a snake. General Reference, Natural History.

Sievert, G., *see* 888

1303 **Sievert, L.** 1987. Captive maintenance of *Crotaphytus collaris.* Bull. Okla. Herp. Soc. 12(1-4):17-18.—Husbandry. *Crotaphytus.*

1304 **Simpson, L. C.** 1974. Paleoecology of the east Manitou Site, southwestern Oklahoma. Okla. Geol. Notes 34(1):15-27.—Fossil amphibians and reptiles included. General Reference, Paleontology.

1305 **Simpson, L. C.** 1976. Paleontology of the Garber Formation (Lower Permian), Tillman County, Oklahoma. M.S. Thesis, Univ. Okla., Norman. 216 pp.—Both fossil amphibians and reptiles included. General Reference, Paleontology.

1306 **Simpson, L. C.** 1976. Paleontology of the Garber Formation (Lower Permian), Tillman County, Oklahoma. Okla. Geol. Notes 36(3):131. —Fossil amphibians and reptiles listed. General Reference, Paleontology.

1307 **Simpson, L. C.** 1977. Bibliography of Permian terrestrial vertebrates of Oklahoma. Compass 54(4):117-120.—Bibliographies, General Reference, Paleontology.

1308 **Simpson, L. C.** 1979. Upper Gearyan and Lower Leonardian terrestrial vertebrate faunas of Oklahoma. Okla. Geol. Notes 39(1):3-21.—Lists Oklahoma sites for fossil amphibians and reptiles. General Reference, Paleontology.

1309 **Sisk, M. E., and C. J. McCoy, Jr.** 1963. Stomach contents of *Natrix r. rhombifera* (Reptilia: Serpentes) from an Oklahoma lake. Proc. Okla. Acad. Sci. 44:68-71.—Natural History. *Nerodia.*

1310 **Sitgreaves, L., and I. C. Woodruff.** 1858. Reports of Captains Sitgreaves and Woodruff of the survey of the Creek Indian boundary line. House Exec. Doc. 104, (12):1-32.—Mentions "reptiles" in report. General Reference, Historical.

1311 **Sizemore, P. D.** 1936. *Cercaria concavocorpa* and n. sp. note on *Hermonimus chelydrae* MacCallum. Trans. Am. Microsc. Soc. 55(4):483-487.—Parasitology. *Chelydra.*

1312 **Sizemore, P. D.** 1936. *Cercaria concavocorpa* n. sp. and note on *Hermonimus chelydrae* MacCallum. M.S. Thesis, Okla. A.M. Coll., Stillwater.—Only available as a bound reprint from the Transactions of the American Microscopy Society. Parasitology. *Chelydra.*

Slater, J. L., *see* 1332

1313 **Smith, A. G.** 1949. The subspecies of the Plains garter snake, *Thamnophis radix.* Bull. Chic. Acad. Sci. 8(14):285-300.—Distribution, Morphology, Taxonomy. *Thamnophis.*

1314 **Smith, C. C.** 1939. Biotic and physiographic succession on abandoned eroded farmland. Ph.D. Diss., Univ. Okla., Norman. 214 pp.—Ecology. *Terrapene, Crotaphytus, Holbrookia, Cnemidophorus, Pituophis.*

1315 **Smith, C. C.** 1940. Biotic and physiographic succession on abandoned eroded farmland. Ecol. Monogr. 10(3):421-484.—Ecology. *Terrapene, Crotaphytus, Holbrookia, Cnemidophorus, Pituophis.*

1316 **Smith, C. C., and G. Acker.** 1939. A collection of reptiles from Logan County, Oklahoma. Proc. Okla. Acad. Sci. 20:65.—Distribution, Natural History. *Terrapene, Crotaphytus, Holbrookia, Phrynosoma, Sceloporus, Cnemidophorus, Diadophis, Elaphe, Masticophis, Nerodia, Pituophis, Sonora, Tantilla.*

1317 **Smith, C. C., and A. N. Bragg.** 1941. Food of *Bufo a. americanus, B. cognatus* and *B. woodhousei* in Oklahoma. Bull. Ecol. Soc. Am. 22(4): 30.—An abstract. Natural History. *Bufo.*

1318 **Smith, C. C., and A. N. Bragg.** 1941. Food of toads in Oklahoma. Anat. Rec. 81(4):111.—An abstract. Natural History. *Bufo.*

1319 **Smith, C. C., and A. N. Bragg.** 1941. Types and quantities of food eaten by four species of Oklahoma toads. Anat. Rec. 81(4):53.—An abstract. Behavior. *Bufo*

1320 **Smith, C. C., and A. N. Bragg.** 1949. Observations on the ecology and natural history of Anura. VII. Food and feeding habits of the common species of toads in Oklahoma. Ecology 30(3):333-349.—Ecology, Natural History. *Bufo.*

Smith, C. C., *see* 378, 379, 380, 381, 382

Smith, D. M., *see* 1339

1321 **Smith, E. N.** 1977. Alligators in Oklahoma. Bull. Okla. Herp. Soc. 2(2-3):35-36.—Distribution, Management, Natural History, Techniques. *Alligator.*

1322 **Smith, G. N.** 1927. The Permian vertebrates of Oklahoma. M.S. Thesis, Univ. Okla., Norman. 46 pp.—Both fossil amphibians and reptiles included. General Reference, Paleontology.

1323 **Smith, H. M.** 1933. On the proper name for the brevicipetid frog *Gastrophryne texensis* (Girard). Copeia 1933(4):217.—Specimens were collected from Lexington in 1932 by H. H. Smith and E. H. Taylor. Study compares the skulls from different locations. Distribution, Taxonomy. *Gastrophryne.*

1324 **Smith, H. M.** 1937. Notes on *Scaphiopus hurterii* Strecker. Herpetologica 1(4):104-108.—Distribution, Taxonomy. *Scaphiopus.*

1325 **Smith, H. M.** 1938. Remarks on the status of the subspecies of *Sceloporus undulatus,* with descriptions of new species and subspecies of the *undulatus* group. Occas. Pap. Mus. Zool. Univ. Mich. (387):1-17.—Distribution, Taxonomy. *Sceloporus.*

1326 **Smith, H. M.** 1946. Handbook of lizards: Lizards of the United States and of Canada. Comstock Publ. Co., Ithaca. 557 pp.—Distribution. *Ophisaurus, Anolis, Crotaphytus, Holbrookia, Phrynosoma, Sceloporus, Eumeces, Scincella, Cnemidophorus.*

1327 **Smith, H. M.** 1946. The systematic status of *Eumeces pluvialis* Cope, and noteworthy records of other amphibians and reptiles from Kansas and Oklahoma. Univ. Kans. Publ. Mus. Nat. Hist. 1:85-89.—Distribution. *Hyla, Eurycea.*

1328 **Smith, H. M.** 1947. Subspecies of the Sonoran toad *(Bufo compactilis* Wiegmann). Herpetologica 4(1):7-13.—Distribution, Taxonomy. *Bufo.*

1329 **Smith, H. M.** 1978. A guide to field identification: Amphibians of North America. Golden Press, New York. 160 pp.—Distribution, Morphology, Natural History, Popular, Taxonomy. All Amphibia.

1330 **Smith, H. M., and E. D. Brodie, Jr.** 1982. A guide to field identification: Reptiles of North America. Golden Press, New York. 240 pp.—Distribution, Morphology, Natural History, Popular, Taxonomy. All Reptilia.

1331 **Smith, H. M., and A. B. Leonard.** 1934. Distributional records of reptiles and amphibians in Oklahoma. Am. Midl. Nat. 15(2):190-196.—Distribution. *Bufo, Acris, Pseudacris, Gastrophryne, Scaphiopus, Rana, Chelydra, Terrapene, Kinosternon, Ophisaurus, Crotaphytus, Holbrookia, Phrynosoma, Sceloporus, Eumeces, Scincella, Cnemidophorus, Arizona, Coluber, Diadophis, Elaphe, Heterodon, Lampropeltis, Masti-*

cophis, Nerodia, Opheodrys, Pituophis, Regina, Rhinocheilus, Sonora, Tantilla, Thamnophis, Tropidoclonion, Virginia.

1332 **Smith, H. M., and J. L. Slater.** 1949. The southern races of *Eumeces septentrionalis* (Baird). Trans. Kans. Acad. Sci. 52(4):438-448.—Distribution, Taxonomy. *Eumeces.*

Smith, H. M., *see* 411, 865, 1228, 1256, 1340, 1341

1333 **Smith, L.** 1979. Strays and accidentals. Outdoor Okla. 35(3):10-13.—Popular. *Alligator.*

1334 **Smith, P. W.** 1957. An analysis of post-Wisconsin biogeography of the prairie peninsula region based on distributional phenomena among terrestrial vertebrate populations. Ecology 38(2):205-218.—Distribution. *Pseudacris, Gastrophryne, Rana, Ambystoma, Plethodon, Siren, Notophthalmus, Terrapene, Kinosternon, Eumeces, Carphophis, Cemophora, Heterodon, Lampropeltis, Pituophis, Thamnophis, Tropidoclonion, Agkistrodon, Crotalus.*

1335 **Smith, P. W.** 1963. *Plethodon cinereus.* Cat. Am. Amphib. Reptiles 5.1-5.3.—Range map and literature review. Bibliographies, Distribution. *Plethodon.*

1336 **Smith, P. W.** 1966. *Hyla avivoca.* Cat. Am. Amphib. Reptiles 28.1-28.2.—Range map and literature review. Bibliographies, Distribution. *Hyla.*

1337 **Smith, P. W.** 1966. *Pseudacris streckeri.* Cat. Am. Amphib. Reptiles 27.1-27.2.—Range map and literature review. Bibliographies, Distribution. *Pseudacris.*

1338 **Smith, P. W.** 1971. Review of R. G. Webb, Reptiles of Oklahoma. Southwest. Nat. 15(4):505-506.—Popular. *Crotaphytus.*

1339 **Smith, P. W., and D. M. Smith.** 1952. The relationship of the chorus frog *Pseudacris nigrita feriarum* and *Pseudacris n. triseriata.* Am. Midl. Nat. 48(1):165-180.—Some of the specimens were from the University of Oklahoma Museum of Zoology. Distribution, Taxonomy. *Pseudacris.*

1340 **Smith, P. W., and H. M. Smith.** 1952. Geographic variation in the lizard *Eumeces anthracinus.* Univ. Kans. Sci. Bull. 34(11):679-694.—Distribution, Taxonomy. *Eumeces.*

1341 **Smith, P. W., and H. M. Smith.** 1962. The systematic and biogeographic status of two Illinois snakes. Occas. Pap. Adams Cent.

Ecol. Stud. (5):1-10.—Article includes the biogeographical significance of Oklahoma. Distribution. *Heterodon, Tropidoclonion.*

Smith, P. W., *see* 411, 1263, 1264

1342 **Snow, J. E.** 1984. Feeding ecology of juvenile turtles (*Chrysemys*) under experimental conditions. Ph.D. Diss., Univ. Okla., Norman. 109 pp.—Behavior, Ecology, Techniques. *Chrysemys, Trachemys.*

1343 **Sobel, L.** 1928. Snakes and mountain boomers. Univ. Okla. Mag.—Lore, Popular. *Ophisaurus, Crotaphytus, Pituophis, Agkistrodon, Crotalus.*

1344 **Soltesz, K.** 1977. Geographical distribution: *Arizona elegans elegans.* Herpetol. Rev. 8(1):13.—Distribution. *Arizona.*

1345 **Soltesz, K.** 1977. Geographical distribution: *Tropidoclonion lineatum.* Herpetol. Rev. 8(1):14.—Distribution. *Tropidoclonion.*

1346 **Spohn, R. T., and S. I. Guttman.** 1976. An electrophoretic study of inter- and intrapopulation genetic variation within the northern fence swift, *Sceloporus undulatus hyacinthinus.* Comp. Biochem. Physiol. 55(B):471-474.—Genetics. *Sceloporus.*

1347 **Spotila, J. R.** 1972. Role of temperature and water in the ecology of lungless salamanders. Ecol. Monogr. 42(1):95-125.—Ecology, Physiology. *Plethodon.*

Springer, M. D., *see* 792

1348 **Stebbins, R. C.** 1951. Amphibians of western North America. Univ. Calif. Press, Berkeley. 359 pp.—Distribution. *Bufo, Gastrophryne, Scaphiopus.*

1349 **Stebbins, R. C.** 1954. Amphibians and reptiles of western North America. McGraw-Hill Book Co., Inc., New York. 528 pp.—Distribution. *Bufo, Acris, Gastrophryne, Scaphiopus, Rana, Ambystoma, Chelydra, Chrysemys, Trachemys, Terrapene, Kinosternon, Trionyx, Crotaphytus, Holbrookia, Phrynosoma, Sceloporus, Eumeces, Cnemidophorus, Arizona, Coluber, Diadophis, Elaphe, Heterodon, Hypsiglena, Lampropeltis, Masticophis, Nerodia, Pituophis, Rhinocheilus, Sonora, Tantilla, Thamnophis, Leptotyphlops, Crotalus, Sistrurus.*

1350 **Stebbins, R. C.** 1966. A field guide to western reptiles and amphibians. Houghton Mifflin Co., Boston. 279 pp.—Distribution, Morphology, Natural History, Popular, Taxonomy. *Bufo, Acris, Pseudacris, Scaphiopus, Rana, Ambystoma, Chelydra, Chrysemys, Trachemys, Terrapene, Kinosternon, Trionyx, Crotaphytus, Holbrookia, Phrynosoma,*

Sceloporus, Uta, Eumeces, Cnemidophorus, Arizona, Coluber, Elaphe, Heterodon, Hypsiglena, Lampropeltis, Masticophis, Nerodia, Opheodrys, Pituophis, Rhinocheilus, Sonora, Storeria, Tantilla, Thamnophis, Tropidoclonion, Leptotyphlops, Crotalus, Sistrurus.

1351 **Steelman, G. M.** 1939. A new cestode from the Texas horned lizard. Trans. Am. Microsc. Soc. 58(4):479-482.—Parasitology. *Phrynosoma.*

1352 **Steelman, G. M.** 1939. *Oocharistica whitentoni,* a new anoplocephalid cestode from a land tortoise. J. Parasitol. 25(6):479-482.—Parasites were from the small intestine of a eastern box turtle collected in Stillwater. Parasitology. *Terrapene.*

1353 **Stejneger, L.** 1895. The poisonous snakes of North America. Ann. Rep. U.S. Nat. Mus. 1893(2):337-487.—Behavior, Development, Distribution, Evolution, Health, Historical, Lore, Morphology, Natural History, Physiology, Taxonomy. *Agkistrodon, Crotalus, Sistrurus.*

1354 **Stejneger, L.** 1938. Restitution of the name *Ptychemys hoyi* Agassiz for a western river tortoise. Proc. Biol. Soc. Wash. (51):173-176.—Distribution, Taxonomy. *Trachemys.*

1355 **Stejneger, L., and T. Barbour.** 1917. A checklist of North American amphibians and reptiles. Harvard Univ. Press, Cambridge. 125 pp.—Distribution. *Eurycea, Notophthalmus, Graptemys, Terrapene, Trionyx, Crotaphytus, Sceloporus, Eumeces, Cnemidophorus, Arizona, Diadophis, Elaphe, Nerodia, Thamnophis, Tropidoclonion, Sistrurus.*

1356 **Stejneger, L., and T. Barbour.** 1923. A checklist of North American amphibians and reptiles. 2nd ed. Harvard Univ. Press, Cambridge. 171 pp.—Distribution. *Eurycea, Notophthalmus, Graptemys, Terrapene, Trionyx, Crotaphytus, Sceloporus, Eumeces, Cnemidophorus, Arizona, Diadophis, Elaphe, Nerodia, Thamnophis, Tropidoclonion, Virginia, Sistrurus.*

1357 **Stejneger, L., and T. Barbour.** 1933. A checklist of North American amphibians and reptiles. 3rd ed. Harvard Univ. Press, Cambridge. 185 pp.—Distribution. *Rana, Eurycea, Notophthalmus, Deirochelys, Graptemys, Terrapene, Kinosternon, Sternotherus, Trionyx, Crotaphytus, Sceloporus, Eumeces, Scincella, Cnemidophorus, Arizona, Carphophis, Diadophis, Elaphe, Lampropeltis, Nerodia, Thamnophis, Tropidoclonion, Virginia, Leptotyphlops, Sistrurus.*

1358 **Stejneger, L., and T. Barbour.** 1939. A checklist of North American amphibians and reptiles. 4th ed. Harvard Univ. Press, Cambridge. 207 pp.—Distribution, Taxonomy. *Bufo, Pseudacris, Gastrophryne,*

Scaphiopus, Rana, Ambystoma, Desmognathus, Eurycea, Plethodon, Typhlotriton, Necturus, Notophthalmus, Chrysemys, Deirochelys, Graptemys, Pseudemys, Trachemys, Terrapene, Kinosternon, Sternotherus, Trionyx, Crotaphytus, Holbrookia, Phrynosoma, Sceloporus, Eumeces, Scincella, Cnemidophorus, All Serpentes.

1359 **Stejneger, L., and T. Barbour.** 1943. A checklist of North American amphibians and reptiles. 5th ed. Bull. Mus. Comp. Zool. 93:1-260. —Distribution, Taxonomy. *Bufo, Pseudacris, Gastrophryne, Scaphiopus, Rana, Ambystoma, Desmognathus, Eurycea, Plethodon, Typhlotriton, Necturus, Notophthalmus, Chrysemys, Deirochelys, Graptemys, Pseudemys, Trachemys, Terrapene, Kinosternon, Sternotherus, Trionyx, Crotaphytus, Holbrookia, Phrynosoma, Sceloporus, Eumeces, Scincella, Cnemidophorus,* All Serpentes, *Alligator.*

1360 **Stewart, J. R., and R. E. Castillo.** 1984. Nutritional provision of the yolk of two species of viviparous reptiles. Physiol. Zool. 57(4):377-383.—Examination of yolk composition in viviparous reptiles. Physiology. *Nerodia.*

1361 **Stewart, J. R., and D. Duvall.** 1985. Ovariectomy fails to block egg-brooding behavior in nesting, female five-lined skinks (*Eumeces fasciatus*). Copeia 1985(3):777-779.—Behavior, Techniques. *Eumeces.*

1362 **Stewart, M. M.** 1983. *Rana clamitans.* Cat. Am. Amphib. Reptiles 337.1-337.4.—Range map and literature review. Bibliographies, Distribution. *Rana.*

1363 **Stewart, M. M., and L. F. Fiuso.** 1982. A bibliography of the green frog, *Rana clamitans* Latreille 1801-1981. Smithson. Herpetol. Infor. Serv. (56):1-53.—Bibliographies. *Rana.*

1364 **Stickel, W. H.** 1938. The snakes of the genus *Sonora* in the United States and lower California. Copeia 1938(4):182-190.—Distribution, Taxonomy. *Sonora.*

1365 **Stone, W.** 1903. A collection of reptiles and batrachians from Arkansas, Indian Territory and western Texas. Proc. Acad. Nat. Sci. Phila. 55:538-542.—Distribution, Historical. *Terrapene, Crotaphytus, Sceloporus, Eumeces, Scincella, Carphophis, Diadophis, Lampropeltis, Nerodia, Tantilla, Agkistrodon.*

1366 **Story, J.** 1958. Okla. Wildl. 14(9):20.—Photograph of diamondback rattlesnake. Popular. *Crotalus.*

1367 **Stott, K.** 1981. Researcher strikes at snake phobia. Bull. Okla. Herp. Soc. 6(2):59.—Reprinted from the Daily Oklahoman. General Reference, Popular.

1368 **Stovall, J. W.** 1932. The Jurassic in Oklahoma. Science 76(1962): 122-123.—The first Jurassic dinosaur from Oklahoma is reported. General Reference, Paleontology.

1369 **Stovall, J. W.** 1935. *Cotylorhynchus romeri,* a new genus and species of pelycosaurian reptile from Oklahoma. Am. J. Sci. 34(202):308-313. —General Reference, Paleontology.

1370 **Stovall, J. W.** 1937. Advance notes on the geology of the Cimarron Valley of western Oklahoma. Proc. Okla. Acad. Sci. 17:78-79.—General Reference, Paleontology.

1371 **Stovall, J. W.** 1937. Geology of the Cimarron River Valley in Cimarron County, Oklahoma. Ph.D. Diss., Univ. Chicago. 60 pp.—General Reference, Paleontology.

1372 **Stovall, J. W.** 1938. The Morrison of Oklahoma and its dinosaurs. J. Geol. 46(4):583-600.—General Reference, Paleontology.

1373 **Stovall, J. W.** 1943. Stratigraphy of the Cimarron Valley (Mesozoic rocks). *In* S. L. Schoff (Ed.). Geology and groundwater resources of Cimarron County, Oklahoma. Okla. Geol. Surv. Bull. (64):43-100.—General Reference, Paleontology.

1374 **Stovall, J. W.** 1948. A new species of embalomerous amphibian from the Permian of Oklahoma. J. Geol. 56(1):75-79.—General Reference, Paleontology.

1375 **Stovall, J. W.** 1950. A new cotylosaur from north central Oklahoma. Am. J. Sci. 248(1):46-54.—A Permian reptile. General Reference, Paleontology.

1376 **Stovall, J. W.** 1952. Oklahoma dinosaurs. Miner. Notes News (Jan).—General Reference, Paleontology, Popular.

1377 **Stovall, J. W., and W. Langston, Jr.** 1950. *Acrocanthosaurus atokensis,* a new genus and species of Lower Cretaceous Theropoda from Oklahoma. Am. Midl. Nat. 43(3):696-728.—General Reference, Paleontology.

1378 **Stovall, J. W., L. I. Price, and A. S. Romer.** 1966. The postcranial skeleton of the giant Permian pelycosaur, *Cotylorhynchus romeri.* Bull. Mus. Comp. Zool. 135(1):1-30.—General Reference, Paleontology.

1379 **Stovall, R. H.** 1982. Applications of micromorphology to natricine systematics (Serpentes, Colubridae): Visual cells and scales. Ph.D. Diss., Okla. State Univ., Stillwater. 162 pp.—Morphology, Taxonomy, Techniques. *Farancia, Nerodia, Regina, Storeria, Thamnophis, Tropidoclonion, Agkistrodon.*

Strecker, J. K., *see* 812

1380 **Stromp, B.** 1981. Voices of the night. Outdoor Okla. 37(2):28-33.—Popular. *Bufo, Acris, Hyla, Pseudacris, Rana.*

1381 **Stromp, B.** 1984. Nature's protection. Outdoor Okla. 40(4):26-33.—Popular. *Terrapene, Phrynosoma.*

1382 **Stull, O. G.** 1932. An annotated list of the forms of the genus *Pituophis.* Occas. Pap. Mus. Zool. Univ. Mich. (250):1-5.—Distribution, Taxonomy. *Pituophis.*

1383 **Stull, O. G.** 1940. Variations and relationships in the snakes of the genus *Pituophis.* Bull. U.S. Natl. Mus. (175):1-225.—Behavior, Bibliographies, Distribution, Historical, Natural History, Taxonomy. *Pituophis.*

1384 **Sturgis, R. S.** 1939. The Wichita Mountains Wildlife Refuge. Chic. Nat. 2(1):9-20.—Distribution. *Crotaphytus, Sceloporus, Eumeces, Cnemidophorus, Diadophis, Elaphe, Masticophis, Nerodia, Pituophis, Thamnophis, Leptotyphlops, Agkistrodon, Crotalus.*

1385 **Sutton, G. M.** 1965. The Black Mesa country of Oklahoma. Pages 291-303 *in* O. S. Pettingill, Jr. (Ed.). The bird watcher's America. 441 pp.—Racerunners and rattlesnakes are mentioned. Natural History. *Cnemidophorus, Crotalus.*

Swenson, F. E., *see* 402

1386 **Taft, C. W.** 1931. Spermatogenesis and seasonal changes in the testes of *Acris gryllus.* Proc. Okla. Acad. Sci. 11:14-15.—Physiology. *Acris.*

Talbot, W. A., *see* 478

1387 **Talley, H. J.** 1937. The annual ovarian cycle of *Acris gryllus.* M.S. Thesis, Univ. Okla., Norman. 51 pp.—Development. *Acris.*

1388 **Tanner, V. M.** 1939. A study of the genus *Scaphiopus.* Great Basin Nat. 1(1):3-26.—Distribution, Evolution, Natural History. *Scaphiopus.*

1389 **Tanner, W. W.** 1944. A taxonomic study of the genus *Hypsiglena.* Great Basin Nat. 5(3-4):25-92.—Distribution, Morphology, Taxonomy. *Hypsiglena.*

1390 **Tanner, W. W., and R. B. Loomis.** 1957. A taxonomic and distributional study of the western subspecies of the milk snake, *Lampropeltis doliata.* Trans. Kans. Acad. Sci. 60(1):12-42.—Distribution, Taxonomy. *Lampropeltis.*

1391 **Tate, R. C.** 1923. Some birds of the Oklahoma Panhandle. Proc. Okla. Acad. Sci. 3:41-51.—Roadrunner observed killing a rattlesnake. Natural History. *Crotalus.*

1392 **Tate, R. C.** 1924. Favorite foods of some Oklahoma birds. Proc. Okla. Acad. Sci. 4:33-35.—Northern harriers and roadrunners were observed eating lizards and snakes. General Reference, Natural History.

1393 **Tate, R. C.** 1925. Some materials used in nest construction of certain birds of the Oklahoma Panhandle. Proc. Okla. Acad. Sci. 5:103-104.—Birds use snake skins in nest. General Reference, Natural History.

1394 **Tayler, B.** 1980. Washington County, Oklahoma, reptile records. Bull. Okla. Herp. Soc. 5(2):23.—Distribution. *Chelydra, Elaphe, Nerodia, Pituophis, Regina, Storeria, Tropidoclonion, Virginia.*

Taylor, D. R., *see* 478

1395 **Taylor, E. H.** 1932. *Eumeces laticeps*: A neglected species of skink. Univ. Kans. Sci. Bull. 20(14):263-271.—Morphology, Taxonomy. *Eumeces.*

1396 **Taylor, E. H.** 1936. Notes and comments on certain American and Mexican snakes of the genus *Tantilla* with descriptions of new species. Trans. Kans. Acad. Sci. 39:335-348.—Distribution, Taxonomy. *Tantilla.*

1397 **Taylor, E. H.** 1936. A taxonomic study of the cosmopolitan scincoid lizards of the genus *Eumeces* with an account of the distribution and relationships of its species. Univ. Kans. Sci. Bull. 36(14):1-643.—Behavior, Bibliographies, Distribution, Ecology, Evolution, Morphology, Taxonomy. *Eumeces.*

Taylor, N., *see* 383

1398 **Taylor, R. J.** 1964. New locations for *Gastrophryne carolinensis* and *Rana clamitans* in Oklahoma. Proc. Okla. Acad. Sci. 45:79-80.—Distribution. *Gastrophryne, Rana.*

1399 **Taylor, R. J., and H. E. Laughlin.** 1964. Additions to the herpetofauna of Bryan County, Oklahoma. Southwest. Nat. 9(1):41-43.—Distribution, Ecology. *Hyla, Rana, Anolis.*

Taylor, R. J., *see* 105, 383

1400 **Taylor, S. R.** 1984. Further information on the natural diet of prairie ringneck snake, *Diadophis punctatus arnyi.* Bull. Okla. Herp. Soc. 9(3):26.—Natural History. *Diadophis.*

1401 **Taylor, W. E.** 1895. The box tortoises of North America. Proc. U.S. Nat. Mus. 17(1019):573-588.—Distribution, Taxonomy. *Terrapene.*

1402 **Thompson, J. N., Jr., and E. M. Grigsby.** 1971. The pickerel frog in northeastern Oklahoma. Southwest. Nat. 16(2):219-220.—Distribution. *Rana, Eurycea.*

1403 **Thurow, G. R.** 1957. Relationships of the red-backed and zig-zag plethodons in the west. Herpetologica 13(2):91-99.—Distribution, Morphology. *Plethodon.*

1404 **Thurow, G. R.** 1966. *Plethodon dorsalis.* Cat. Am. Amphib. Reptiles 29.1-29.2.—Range map and literature review. Bibliographies, Distribution. *Plethodon.*

1405 **Tihen, J. A.** 1955. A new Pliocene species of *Ambystoma,* with remarks on other fossil ambystomids. Contrib. Mus. Paleontol. Univ. Mich. 12(11):229-244.—Paleontology. *Ambystoma.*

1406 **Tihen, J. A.** 1958. Comments on the osteology and phylogeny of ambystomatid salamanders. Bull. Fla. State Mus. Biol. Sci. 3:1-50.—Distribution, Morphology, Taxonomy. *Ambystoma.*

1407 **Tihen, J. A.** 1960. Notes on Late Cenozoic hylid and leptodactylid frogs from Kansas, Oklahoma, and Texas. Southwest. Nat. 5(2):66-70.—Paleontology. *Pseudacris.*

1408 **Tihen, J. A.** 1962. Osteological observations on new world *Bufo.* Am. Midl. Nat. 67(1):157-183.—Morphology. *Bufo.*

1409 **Tihen, J. A.** 1962. A review of new world fossil bufonids. Am. Midl. Nat. 68(1):1-50.—Morphology, Paleontology. *Bufo.*

1410 **Tihen, J. A.** 1969. *Ambystoma.* Cat. Am. Amphib. Reptiles 75.1-75.4.—Range map and literature review. Bibliographies, Distribution. *Ambystoma.*

1411 **Tinkle, D. W.** 1958. The systematics and ecology of the *Sternotherus carinatus* complex (Testudinata, Chelydridae). Tulane Stud. Zool. 6(1):1-56.—Distribution, Morphology, Natural History, Taxonomy. *Sternotherus.*

1412 **Tomer, J. S.** 1985. Edith Force Kassing: Scientist with a gift for teaching. Chron. Okla. 63(4):396-411.—Historical. *Rana.*

1413 **Trainer, J. E.** 1971. An ultrastructure investigation of body wall components in *Procephalus crotali* (Pentastomida). Ph.D. Diss., Univ. Okla., Norman. 78 pp.—Parasitology. *Crotalus.*

1414 **Trainer, J. E., J. T. Self, and K. E. Richter.** 1974. An ultra-structure study of the Pentastome cuticle. J. Parasitol. 61(4):753-758.—Parasitology. *Crotalus.*

1415 **Trapido, H.** 1939. Studies on the taxonomy and natural history of the snakes of the genus *Storeria* in the United States and Canada. M.S. Thesis, Cornell Univ., Ithaca, New York. 100 pp.—Oklahoma specimens used. Natural History, Taxonomy. *Storeria.*

1416 **Trapido, H.** 1944. The snakes of the genus *Storeria.* Am. Midl. Nat. 31(1):1-84.—Distribution, Evolution, Taxonomy. *Storeria.*

1417 **Trapp, M. M.** 1956. Range and natural history of the ringed salamander, *Ambystoma annulatum* Cope (Ambystomidae). Southwest. Nat. 1(2):78-82.—Behavior, Distribution, Natural History. *Ambystoma.*

1418 **Trauth, S. E.** 1983. *Lampropeltis calligaster calligaster* (prairie kingsnake). Predation. Herpetol. Rev. 14(3):74.—Behavior. *Lampropeltis.*

1419 **Trauth, S. E.** 1983. Nesting habits and reproductive characteristics of the lizard *Cnemidophorus sexlineatus* (Lacertilia: Teiidae). Am. Midl. Nat. 109(2):289-299.—Natural History. *Cnemidophorus.*

1420 **Trowbridge, A. H.** 1937. Ecological observations on amphibians and reptiles collected in southeastern Oklahoma during the summer of 1934. Am. Midl. Nat. 18(2):285-303.—Distribution, Ecology, Historical. *Bufo, Acris, Hyla, Pseudacris, Gastrophryne, Rana, Ambystoma, Eurycea, Hemidactylium, Plethodon, Siren, Notophthalmus, Chelydra, Macroclemys, Deirochelys, Graptemys, Pseudemys, Trachemys, Terrapene, Sternotherus, Trionyx, Ophisaurus, Anolis, Crotaphytus, Scelop-*

orus, Eumeces, Scincella, Cnemidophorus, Carphophis, Coluber, Diadophis, Elaphe, Heterodon, Lampropeltis, Masticophis, Nerodia, Opheodrys, Storeria, Tantilla, Thamnophis, Agkistrodon, Crotalus.

1421 **Trowbridge, A. H.** 1937. New records of Amphibia for Oklahoma. Copeia 1937(1):71-72.—Distribution. *Hyla, Pseudacris, Rana, Ambystoma, Desmognathus, Siren.*

1422 **Trowbridge, A. H., and H. M. Hefley.** 1933. Preliminary studies on the parasite fauna of Oklahoma anurans. Proc. Okla. Acad. Sci. 14: 16-19.—Parasitology. *Bufo, Acris, Rana.*

1423 **Trowbridge, A. H., and M. S. Trowbridge.** 1937. Notes on the cleavage rate of *Scaphiopus bombifrons* Cope, with additional remarks on certain aspects of its life history. Am. Nat. 71(736):460-480.—Behavior, Development, Natural History. *Bufo, Scaphiopus.*

1424 **Trowbridge, M. S.** 1939. Embryonic and larval development of *Scaphiopus bombifrons* Cope. Ph.D. Diss., Univ. Okla., Norman. 159 pp.—Development, Natural History. *Scaphiopus.*

1425 **Trowbridge, M. S.** 1941. Studies on the normal development of *Scaphiopus bombifrons.* I. The cleavage period. Trans. Am. Microsc. Soc. 60(4):508-526.—Development. *Scaphiopus.*

1426 **Trowbridge, M. S.** 1942. Studies on the normal development of *Scaphiopus bombifrons.* II. The later embryonic and larval periods. Trans. Am. Microsc. Soc. 61(1):66-83.—Development. *Scaphiopus.*

Trowbridge, M. S., *see* 1423

Tucker, J. K., *see* 668

1427 **Turtlet, L. A.** 1924. The preservation of wildlife. Proc. Okla. Acad. Sci. 4:35-42.—Article suggests that rattlesnakes should be destroyed to save quail from predation. Natural History. *Crotalus.*

1428 **Twente, J. W., Jr.** 1952. Pliocene lizards from Kansas. Copeia 1952-(2):70-73.—Compares fossil *Sceloporus* teeth from Kansas to recent *Sceloporus* from Oklahoma. Morphology, Paleontology. *Sceloporus.*

1429 **Twente, J. W., Jr.** 1955. Aspects of a population study of cavern-dwelling bats. J. Mammal. 36(3):379-390.—A rat snake was preying on bats. Ecology. *Elaphe.*

1430 **Tyler, J. D.** 1969. Distribution and vertebrate associates of the black-tailed prairie dog in Oklahoma. Ph.D. Diss., Univ. Okla., Norman. 85 pp.—Distribution, Ecology. *Bufo, Gastrophryne, Scaphiopus,*

Ambystoma, Terrapene, Kinosternon, Holbrookia, Phrynosoma, Sceloporus, Cnemidophorus, Pituophis, Crotalus.

1431 **Tyler, J. D.** 1970. Vertebrates in a prairie dog town. Proc. Okla. Acad. Sci. 50:110-113.—Natural History. *Bufo, Acris, Scaphiopus, Ambystoma, Crotaphytus, Phrynosoma, Cnemidophorus, Pituophis, Rhinocheilus, Sonora, Tantilla, Leptotyphlops.*

1432 **Tyler, J. D.** 1976. Eruption of juvenile bullfrogs in Comanche County, Oklahoma. Bull. Okla. Herp. Soc. 1(4):55.—Natural History. *Rana.*

1433 **Tyler, J. D.** 1977. An "active" bullfrog in late fall. Bull. Okla. Herp. Soc. 2(1):15-16.—Natural History. *Rana.*

1434 **Tyler, J. D.** 1977. Coachwhip preys on horned lizard. Southwest. Nat. 22(1):146.—Natural History. *Masticophis, Phrynosoma.*

1435 **Tyler, J. D.** 1977. Ground skink with bifurcate tail. Bull. Okla. Herp. Soc. 2(2-3):51.—Morphology. *Scincella.*

1436 **Tyler, J. D.** 1978. Bullfrog observed eating Blanchard's cricket frog. Bull. Okla. Herp. Soc. 3(4):68-69.—Natural History. *Acris, Rana.*

1437 **Tyler, J. D.** 1978. Western coachwhip scavenger of eastern collared lizard. Bull. Okla. Herp. Soc. 3(4):73.—Natural History. *Crotaphytus, Masticophis.*

1438 **Tyler, J. D.** 1979. A case of swimming in *Terrapene carolina* (Testudines: Emydidae). Southwest. Nat. 24(1):189-190.—Behavior. *Terrapene.*

1439 **Tyler, J. D.** 1982. Cyanistic bullfrog from southwestern Oklahoma. Southwest. Nat. 27(2):218.—Morphology. *Rana.*

1440 **Tyler, J. D.** 1982. Ground skink with bifurcate tail. Bull. Okla. Herp. Soc. 7(3-4):88-89.—Same article as Tyler (1977) above. Morphology. *Scincella.*

1441 **Tyler, J. D., and J. W. Ault, III.** 1978. Herptiles inhabiting an old cistern. Bull. Okla. Herp. Soc. 3(1):11-12.—Natural History. *Bufo, Pseudacris, Gastrophryne, Scaphiopus, Ambystoma, Pituophis.*

1442 **Tyler, J. D., and H. N. Buscher.** 1980. Notes on a population of larval *Ambystoma tigrinum* (Ambystomatidae) from Cimarron County, Oklahoma. Southwest. Nat. 25(3):391-397.—Ecology, Natural History, Parasitology. *Ambystoma.*

1443 **Tyler, J. D., and M. England.** 1981. Breeding of *Crotalus atrox* in Comanche County, Oklahoma. Bull. Okla. Herp. Soc. 6(2):46.—Behavior, Natural History. *Crotalus.*

1444 **Tyler, J. D., and R. D. Hoestenbach, Jr.** 1979. Differences in food of bullfrogs (*Rana catesbeiana*) from pond and stream habitats in southwestern Oklahoma. Southwest. Nat. 24(1):33-38.—Ecology. *Rana.*

1445 **Tyler, J. D., and L. McGee.** 1976. Albino lined snake (*Tropidoclonion lineatum*) from southwestern Oklahoma. Southwest. Nat. 21(3):403.—Morphology. *Tropidoclonion.*

1446 **Tyler, J. D., and D. C. Parmley.** 1981. Occurrence of *Terrapene carolina triunguis* (Testudinidae) in southwestern Oklahoma. Bull. Okla. Herp. Soc. 6(1):4-5.—Distribution. *Terrapene.*

Tyler, J. D., *see* 1167, 1168

1447 **Vaeth, R. H.** 1980. New herpetological county records for north-central Oklahoma. Bull. Okla. Herp. Soc. 5(3):46-47.—Distribution. *Masticophis, Nerodia, Opheodrys, Leptotyphlops, Agkistrodon, Crotalus.*

1448 **Van Vleet, A. H.** 1902. Snakes of Oklahoma. Pages 167-173 *in* Sec. Bien. Rep. Dept. Geol. Nat. Hist. Terr. Okla.—Distribution. *Carphophis, Coluber, Diadophis, Elaphe, Heterodon, Lampropeltis, Nerodia, Opheodrys, Pituophis, Thamnophis, Agkistrodon, Crotalus, Sistrurus.*

1449 **Vance, T.** 1986. Two unreported turtle records for Love County, Oklahoma. Bull. Okla. Herp. Soc. 11(1-4):13.—Distribution. *Graptemys, Pseudemys.*

1450 **Vanderpol, B.** 1955. Horned toads. Okla. Game Fish News 11(6):inside back cover.—A reader pointed out an error in a previous article; horned lizards do lay eggs. Natural History, Popular. *Phrynosoma.*

1451 **Vaughn, P. P.** 1958. On a new pelycosaur from the Lower Permian of Oklahoma, and on the origin of the family Caseidae. J. Paleontol. 32(5):981-991.—General Reference, Paleontology.

1452 **Vaughn, P. P.** 1958. A specimen of the captorhinid reptile, *Captorhinkas chozaensis* Olson, 1954, from the Hennessey Formation, Lower Permian of Oklahoma. J. Geol. 66(3):327-332.—General Reference, Paleontology.

Vaughn, P. P., *see* 1138

1453 **Vehik, R.** 1982. The archeology of the Bug Hill Site (34Pu-116): Pushmataha County, Oklahoma. Univ. Okla. Archeol. Res. Manage. Cent. Res. Ser. (7):1-226.—Archeology. *Scaphiopus, Rana, Chelydra,*

Graptemys, Trachemys, Terrapene, Kinosternon, Sternotherus, Trionyx, Crotaphytus, Coluber, Crotalus.

Vinegar, A., *see* 818

1454 **Vinyard, W. C.** 1953. Epizoophytic algae from mollusks, turtles, and fish in Oklahoma. Proc. Okla. Acad. Sci. 34:63-65.—Natural History. *Chelydra, Graptemys, Trachemys, Trionyx.*

1455 **Viosca, P. J.** 1937. A tentative revision of the genus *Necturus* with descriptions of three new species from the southern gulf drainage region. Copeia 1937(2):120-138.—Distribution, Natural History, Taxonomy. *Necturus.*

1456 **Vogt, R. C., and C. J. McCoy, Jr.** 1980. Status of the emydine turtle genera *Chrysemys* and *Pseudemys.* Ann. Carnegie Mus. 49(5):93-102.—Genetics, Morphology. *Chrysemys, Pseudemys, Trachemys.*

1457 **Volpe, E. P.** 1953. Embryonic temperature adaptations and relationships in toads. Physiol. Zool. 26(4):344-354.—Development. *Bufo.*

1458 **Volpe, E. P.** 1959. Experimental and natural hybridization between *Bufo terrestris* and *Bufo fowleri.* Am. Midl. Nat. 61(2):295-312.—Shows range in Oklahoma. Distribution. *Bufo.*

1459 **Walker, J. M., M. A. Paulissen, and J. M. Britton.** 1986. Habitat diversity in the whiptail lizard *Cnemidophorus gularis gularis* (Teiidae) in southern Oklahoma. Southwest. Nat. 31(3):405-408.—Behavior, Ecology. *Cnemidophorus.*

1460 **Ward, J. P.** 1978. *Terrapene ornata.* Cat. Am. Amphib. Reptiles 217.1-217.4.—Range map and literature review. Bibliographies, Distribution. *Terrapene.*

1461 **Ward, J. P.** 1984. Relationships of chrysemyid turtles of North America (Testudines: Emydidae). Spec. Publ. Mus. Tex. Tech. Univ. (21):1-50.—Distribution, Taxonomy. *Chrysemys, Pseudemys, Trachemys.*

1462 **Warner, M.** 1977. Herpetol. Rev. 8(3):112.—Photograph of western painted turtle from Grant County, Oklahoma. Distribution, Popular. *Chrysemys.*

1463 **Warren, D.** 1974. Oklahoma's endangered species. Outdoor Okla. 30(8):17-19.—Popular. *Alligator.*

1464 **Warren, D.** 1976. Bagging bulls and boating cats. Outdoor Okla. 32(7):19-20.—Popular. *Rana.*

1465 **Warren, J. W.** 1961. The basicranial articulation of the Early Permian cotylosaur, *Captorhinus*. J. Paleontol. 35(3):561-563.—Fossil reptile from Fort Sill fissure deposit. General Reference, Paleontology.

1466 **Wasserman, A. O.** 1957. Factors affecting interbreeding in sympatric species of spadefoots (*Scaphiopus*). Evolution 11(3):320-338.—Distribution. *Scaphiopus*.

1467 **Wasserman, A. O.** 1968. *Scaphiopus holbrookii*. Cat. Am. Amphib. Reptiles 70.1-70.4.—Range map and literature review. Bibliographies, Distribution. *Scaphiopus*.

1468 **Wasserman, A. O.** 1970. *Scaphiopus couchii*. Cat. Am. Amphib. Reptiles 85.1-85.4.—Range map and literature review. Bibliographies, Distribution. *Scaphiopus*.

Wasserzug, R. J., *see* 223

1469 **Watkins, II., J. F., F. R. Gehlbach, and J. C. Kroll.** 1969. Attractant-repellant secretions of blind snakes (*Leptotyphlops dulcis*) and their army ant prey (*Neivamyrmex nigrescens*). Ecology 50(6):1098-1102.—Specimens were collected from Marshall and Seminole counties. Behavior, Distribution, Ecology, Techniques. *Diadophis, Lampropeltis, Sonora, Tantilla, Leptotyphlops*.

1470 **Watkins, II, J. F., F. R. Gehlbach, and R. W. Plsek.** 1972. Behavior of blind snakes (*Leptotyphlops dulcis*), response to army ant (*Neivamyrmex nigrescens*) raiding columns. Tex. J. Sci. 23(4):556-557.—Some of the snakes were collected in Seminole County. Behavior. *Leptotyphlops*.

Watkins, II, J. F., *see* 679

Webb, O. L., *see* 972

1471 **Webb, R. G.** 1950. Range extension of the chicken turtle in Oklahoma. Herpetologica 6(5):137-138.—Distribution. *Deirochelys*.

1472 **Webb, R. G.** 1952. The reptiles of Oklahoma. M.S. Thesis, Univ. Okla., Norman. 282 pp.—Bibliographies, Distribution, Ecology, Historical, Taxonomy. All Testudines, All Sauria, All Serpentes.

1473 **Webb, R. G.** 1956. Size at sexual maturity in the male softshell turtle, *Trionyx ferox emoryi*. Copeia 1956(2):121-122.—Development, Morphology. *Trionyx*.

1474 **Webb, R. G.** 1961. Observations on the life histories of turtles (Genus *Pseudemys* and *Graptemys*) in Lake Texoma, Oklahoma. Am.

Midl. Nat. 65(1):193-214.—Ecology, Natural History. *Graptemys, Pseudemys, Trachemys.*

1475 **Webb, R. G.** 1962. North American recent soft-shelled turtles (Family Trionychidae). Univ. Kans. Publ. Mus. Nat. Hist. 13(10):429-611.—Bibliographies, Distribution, Ecology, Evolution, Historical, Natural History, Paleontology, Taxonomy. *Trionyx.*

1476 **Webb, R. G.** 1970. Reptiles of Oklahoma. Univ. Okla. Press, Norman, Oklahoma. 370 pp.—Bibliographies, Distribution, Ecology, Historical, Natural History, Taxonomy. All Reptilia.

1477 **Webb, R. G.** 1973. *Trionyx muticus.* Cat. Am. Amphib. Reptiles 139.1-139.2—Range map and literature review. Bibliographies, Distribution. *Trionyx.*

1478 **Webb, R. G.** 1973. *Trionyx spiniferus.* Cat. Am. Amphib. Reptiles 140.1-140.4.—Range map and literature review. Bibliographies, Distribution. *Trionyx.*

1479 **Webb, R. G.** 1980. *Thamnophis cyrtopsis.* Cat. Am. Amphib. Reptiles 245.1-245.4.—Range map and literature review. Bibliographies, Distribution. *Thamnophis.*

1480 **Webb, R. G.** 1981. Indiscriminate rattlesnake killing should be stopped. Bull. Okla. Herp. Soc. 6(1):5-6.—Management. *Crotalus.*

1481 **Webb, R. G., and A. I. Ortenburger.** 1953. Reptiles of the Wichita Mountains Wildlife Refuge, Comanche County, Oklahoma. Proc. Okla. Acad. Sci. 34:87-92.—Distribution. *Chelydra, Graptemys, Trachemys, Terrapene, Kinosternon, Trionyx, Ophisaurus, Crotaphytus, Holbrookia, Phrynosoma, Sceloporus, Eumeces, Scincella, Cnemidophorus, Arizona, Coluber, Diadophis, Elaphe, Heterodon, Hypsiglena, Lampropeltis, Masticophis, Nerodia, Opheodrys, Pituophis, Regina, Rhinocheilus, Sonora, Storeria, Tantilla, Thamnophis, Tropidoclonion, Leptotyphlops, Agkistrodon, Crotalus, Sistrurus.*

1482 **Webb, W. L.** 1950. Biogeographic regions of Texas and Oklahoma. Ecology 31(3):426-433.—The three families of snakes were included. Distribution, Ecology, General Reference.

1483 **Webster, E. B.** 1935. A preliminary list of the reptiles and amphibians of Pottawatomie County, Oklahoma, near Shawnee. Proc. Okla. Acad. Sci. 16:20-22.—Distribution. *Bufo, Acris, Hyla, Rana, Ambystoma, Chelydra, Trachemys, Terrapene, Trionyx, Ophisaurus, Crotaphytus, Phrynosoma, Sceloporus, Cnemidophorus, Carphophis, Coluber, Diadophis, Elaphe, Heterodon, Lampropeltis, Masticophis,*

Nerodia, Opheodrys, Pituophis, Regina, Sonora, Storeria, Tantilla, Thamnophis, Tropidoclonion, Leptotyphlops, Agkistrodon.

Webster, T. P., *see* 775

Weese, A. O., *see* 384, 385, 386

1484 **Wegemar, C. H.** 1915. Anticlinal structure in part of Cotton and Jefferson counties, Oklahoma. U.S. Geol. Surv. Bull. (602):1-108.—General Reference, Paleontology.

Weinstein, S. A., *see* 1062

1485 **Wenglarz, E. D.** 1980. First Grant County, Oklahoma, record of western coachwhip. Bull. Okla. Herp. Soc. 5(1):6.—Distribution. *Elaphe, Masticophis.*

1486 **Wenglarz, E. D.** 1980. Reptiles records for Grant County, Oklahoma. Bull. Okla. Herp. Soc. 4(4):69.—Distribution. *Phrynosoma, Sistrurus.*

1487 **West, E. S.** 1978. Biostratrgraphy and paleoecology of the Lower Morrison Formation, of Cimarron County, Oklahoma. M.S. Thesis, Wichita State Univ., Wichita, Kansas. 61 pp.—General Reference, Paleontology.

1488 **West, L. W., and R. D. Grimpe.** 1977. Suggested procedures to instigate breeding in captive reptiles. Bull. Okla. Herp. Soc. 4(2-3):35-36.—General Reference, Husbandry.

1489 **Whaley, M. C.** 1930. Spermatogenesis and seasonal changes in the testes of *Acris gryllus.* M.S. Thesis, Univ. Okla., Norman. 64 pp.—Development. *Acris.*

Whitehead, G., *see* 161

1490 **Wickham, M. M.** 1922. Further notes on the migration of *Terrapene carolina* in Oklahoma. Proc. Okla. Acad. Sci. 2:22.—Natural History. *Terrapene.*

1491 **Wickham, M. M.** 1922. Notes on the migration of *Macrochelys lacertina.* Proc. Okla. Acad. Sci. 2:20-22.—Natural History. *Macroclemys.*

1492 **Wiens, J.** 1984. Plethodontid salamanders Part 1: Distribution. Colo. Herpetol. Soc. Newsl. 12(4):2-4.—Distribution. *Eurycea, Plethodon, Typhlotriton.*

1493 **Wille, C.** 1974. Conservation notebook: Oklahoma amphibians. Outdoor Okla. 30(7):centerfold.—Popular. *Bufo, Acris, Rana, Eurycea.*

1494 **Wille, C.** 1976. Rattlesnake roundups. Outdoor Okla. 32(7):6-9.—Management, Popular. *Crotalus.*

1495 **Wille, C.** 1977. Conservation notebook: Fur, fin & feather. Outdoor Okla. 33(11):centerfold.—Close up photographs of integument. Popular. *Bufo, Terrapene.*

1496 **Wille, C.** 1977. Conservation notebook: Oklahoma reptiles. Outdoor Okla. 33(4):centerfold.—Popular. *Trachemys, Terrapene, Crotaphytus, Phrynosoma, Lampropeltis, Opheodrys.*

1497 **Williams, J. L.** 1973. Black rat snake nest-predation on wood duck in Oklahoma. Bull. Okla. Ornithol. Soc. 6(2):15.—Natural History. *Elaphe.*

1498 **Williams, K. L.** 1971. Systematics of the colubrid snake *Lampropeltis triangulum* Lacepede. Ph.D. Diss., La. State Univ., Baton Rouge. 458 pp.—Distribution, Natural History, Paleontology, Taxonomy. *Lampropeltis.*

1499 **Williams, K. L.** 1978. Systematics and natural history of the American milk snake, *Lampropeltis traingulum.* Milw. Public Mus. Publ. Biol. Geol. (2):1-258.—Distribution, Natural History, Paleontology, Taxonomy. *Lampropeltis.*

1500 **Williams, K. L.** 1985. *Cemophora, C. coccinea.* Cat. Am. Amphib. Reptiles 374.1-374.4.—Range map and literature review. Bibliographies, Distribution. *Cemophora.*

1501 **Williams, K. L., B. C. Brown, and L. D. Wilson.** 1966. A new subspecies of the colubrid snake *Cemophora coccinea* (Blumenbach) from southern Texas. Tex. J. Sci. 18(1):85-88.—Distribution, Taxonomy. *Cemophora.*

1502 **Williams, K. L., and L. D. Wilson.** 1967. A review of the colubrid snake genus *Cemophora* Cope. Tulane Stud. Zool. 13(4):103-124.—Bibliographies, Distribution, Ecology, Natural History, Taxonomy. *Cemophora.*

1503 **Williams, R. W.** 1952. A survey of the helminths of the snapping turtle, *Chelydra serpentina,* from Oklahoma. M.S. Thesis, Okla. A.M. Coll., Stillwater. 76 pp.—Parasitology. *Chelydra.*

1504 **Williams, R. W.** 1953. Helminths of the snapping turtle, *Chelydra serpentina serpentina,* from Oklahoma; including the first report and description of the male of *Capillaria serpentina* Harwood, 1932. Trans. Am. Microsc. Soc. 72(2):175-178.—Parasitology. *Chelydra.*

1505 **Williams, V.** 1956. Okla. Game Fish News 12(4):4.—Photograph of timber rattlesnake. Popular. *Crotalus.*

1506 **Williston, S. W.** 1899. Notes on the coraco-scapula of *Eryops* (Cope). Kans. Univ. Quart. 8:185-186.—Oklahoma amphibians used. General Reference, Paleontology.

1507 **Wilson, C.** 1974. Oklahoma alligators: Strange but true. Outdoor News (Okla. Wildl. Fed.) 22(5):14.—Distribution, Popular. *Alligator.*

1508 **Wilson, E. E.** 1977. Herp hunting in eastern Oklahoma County. Bull. Okla. Herp. Soc. 2(1):17-18.—Distribution. *Bufo, Acris, Hyla, Pseudacris, Scaphiopus, Rana, Trionyx, Crotaphytus, Phrynosoma, Sceloporus, Eumeces, Scincella, Cnemidophorus, Diadophis, Pituophis, Sonora, Tantilla, Virginia, Leptotyphlops, Agkistrodon, Crotalus.*

1509 **Wilson, L. D.** 1970. The coachwhip snake, *Masticophis flagellum* (Shaw): Taxonomy and distribution. Tulane Stud. Zool. Bot. 16(2): 31-99.—Distribution, Taxonomy. *Masticophis.*

1510 **Wilson, L. D.** 1973. *Masticophis.* Cat. Am. Amphib. Reptiles 144.1-144.2.—Range map and literature review. Bibliographies, Distribution. *Masticophis.*

1511 **Wilson, L. D.** 1973. *Masticophis flagellum.* Cat. Am. Amphib. Reptiles 145.1-145.4—Range map and literature review. Bibliographies, Distribution. *Masticophis.*

1512 **Wilson, L. D.** 1978. *Coluber constrictor.* Cat. Am. Amphib. Reptiles 218.1-218.4.—Range map and literature review. Bibliographies, Distribution. *Coluber.*

1513 **Wilson, L. D.** 1982. *Tantilla.* Cat. Am. Amphib. Reptiles 307.1-307.4.—Range map and literature review. Bibliographies, Distribution. *Tantilla.*

Wilson, L. D., *see* 1265, 1501, 1502

1514 **Woodburne, M. O.** 1956. Notes on the snake *Sistrurus catenatus tergeminus* in southwestern Kansas and northwestern Oklahoma. Copeia 1956(2):125-126.—Distribution. *Sistrurus.*

1515 **Woodburne, M. O.** 1959. A fossil alligator from the Lower Pliocene of Oklahoma and its climatic significance. Pap. Mich. Acad. Sci. Arts Lett. (44):47-51.—Paleontology. *Alligator.*

Woodruff, I. C., *see* 1310

Wright, A. A., *see* 1517, 1518, 1519, 1520, 1521, 1522

1516 **Wright, A. H.** 1929. Synopsis and description of North American tadpoles. Proc. U.S. Nat. Mus. 74(11):1-70.—Morphology. All Anura.

1517 **Wright, A. H., and A. A. Wright.** 1952. List of the snakes of the United States and Canada by states and provinces. Am. Midl. Nat. 48(3):574-603.—Distribution. All Serpentes.

1518 **Wright, A. H., and A. A. Wright.** 1957. Handbook of snakes of the United States and Canada. Volume I. Cornell Univ. Press, Ithaca. 564 pp.—Distribution. *Arizona, Carphophis, Cemophora, Coluber, Diadophis, Elaphe, Farancia, Heterodon, Hypsiglena, Lampropeltis, Masticophis, Nerodia, Opheodrys.*

1519 **Wright, A. H., and A. A. Wright.** 1957. Handbook of snakes of the United States and Canada. Volume II. Cornell Univ. Press, Ithaca. 540 pp.—Distribution. *Pituophis, Regina, Rhinocheilus, Sonora, Storeria, Tantilla, Thamnophis, Tropidoclonion, Virginia, Leptotyphlops, Agkistrodon, Crotalus, Sistrurus.*

1520 **Wright, A. H., and A. A. Wright.** 1962. Handbook of snakes of the United States and Canada. Volume III. Bibliography. Edwards Bros., Inc., Ann Arbor. 179 pp.—Bibliographies. All Serpentes.

1521 **Wright, A. H., and A. A. Wright.** 1970. Handbook of Frogs and Toads of the United States and Canada. 3rd ed. Cornell Univ. Press, Ithaca.—Distribution. All Anura.

1522 **Wright, A. H., and A. A. Wright.** 1979. Handbook of snakes of the United States and Canada. Volume III. Bibliography. Soc. Study Amphib. Reptiles Facs. Reprints Herpetol. 179 pp.—Bibliographies. All Serpentes.

1523 **Wright, C., Jr.** 1981. Future of the O.H.S. Bull. Okla. Herp. Soc. 6(4):83-84.—General Reference, Popular.

Wright, D. A., *see* 779

1524 **Wyckoff, D. G.** 1976. Faunal materials from four early caddoan tradition sites, Arkansas Basin, eastern Oklahoma. Okla. Archeol. Surv. 22 pp.—Unpublished manuscript. Archeology. *Chelydra, Chrysemys, Graptemys, Trachemys, Terrapene, Trionyx.*

1525 **Yarrow, H. C.** 1882. Checklist of North American Reptilia and Batrachia, with Catalog of Specimens in U.S. National Museum. Bull. U.S. Natl. Mus. (24):1-249.—Distribution. *Bufo, Acris, Rana, Ambystoma, Eurycea, Trionyx, Crotaphytus, Phrynosoma, Sceloporus, Eumeces, Cnemidophorus, Cemophora, Coluber, Elaphe, Heterodon, Lampropeltis, Masticophis, Nerodia, Thamnophis, Agkistrodon, Crotalus.*

1526 **Yarrow, H. C.** 1883. Description of new species of reptiles in the United States National Museum. Proc. U.S. Nat. Mus. 6:152-154.—Distribution, Historical. *Masticophis.*

1527 **Yeomans, J. A.** 1977. An evaluation of wildlife enhancement activities in the residential context. M.S. Thesis, Okla. State Univ., Stillwater. 85 pp.—General Reference, Management.

1528 **Yoshida, J. K.** 1966. Studies on the development of social behavior in some hatchling and juvenile iguanid and agamid lizards. M.S. Thesis, Univ. Okla., Norman. 55 pp.—Behavior. *Sceloporus.*

Yoshida, J. K., *see* 481

1529 **Zidek, J.** 1971. *Labidosaurikos meachami,* a Lower Permian reptile. Okla. Geol. Notes 31(5):90.—General Reference, Paleontology.

1530 **Zimmerman, J. W.** 1982. The common long-nosed armadillo (*Dasypus novemcinctus*) in northcentral Oklahoma. M.S. Thesis, Okla. State Univ., Stillwater. 106 pp.—Lizards and snakes comprised 1.8% of the diet. General Reference, Natural History.

1531 **Zimmerman, R. M.** 1952. Intestinal protozoa from some lizards common to Oklahoma. M.S. Thesis, Univ. Okla., Norman. 55 pp.—Parasitology. *Phrynosoma, Sceloporus, Eumeces, Scincella, Cnemidophorus.*

1532 **Zimmerman, R. M., and H. P. Brown.** 1952. Observations on some intestinal protozoa in Oklahoma lizards, with the description of a new genus, *Biflagella.* Proc. Okla. Acad. Sci. 33:103-111.—Parasitology. *Phrynosoma, Sceloporus, Eumeces, Scincella, Cnemidophorus.*

1533 **Zug, G. R.** 1986. *Sternotherus.* Cat. Am. Amphib. Reptiles 397.1-397.3.—Bibliographies, Distribution, Taxonomy. *Sternotherus.*

1534 **Zug, G. R., and A. Schwartz.** 1971. *Deirochelys, D. reticularia.* Cat. Am. Amphib. Reptiles 107.1-107.3.—Range map and literature review. Bibliographies, Distribution. *Deirochelys.*

1535 **Zweifel, R. C.** 1956. Two pelobatid frogs from the Tertiary of North America and their relationships to fossil and recent forms. Am. Mus. Novit. (1762):1-45.—Bibliographies, Distribution, General Reference, Paleontology, Taxonomy.

1536 **Zweifel, R. C.** 1965. Variation in and distribution of the unisexual lizard, *Cnemidophorus tesselatus.* Am. Mus. Novit. (2235):1-49.—Distribution, Taxonomy. *Cnemidophorus.*

Zweifel, R. C., *see* 587

Genus and Key Word Index

Genus and Key Word Index

Listing of sequence numbers for references in main body of bibliography that pertain to particular genera of amphibians or reptiles. Note that those references that cover fossil forms or that do not make specific reference to genera or species are listed under the category of "General Reference" at the end of the index.

BUFONIDAE (BUFONIDS AND TOADS)

Bufo (Toads)

Bufo (Toads) (*continued*)

Bufo (Toads) (*continued*)

HYLIDAE (HYLIDS AND TREEFROGS)

Acris (Cricket frogs)

Acris (Cricket frogs) (*continued*)

Hyla (Treefrogs)

Pseudacris (Chorus frogs) (*continued*)

Popular

47 101 146 1004 1101 1329 1350 1380

Taxonomy

101 274 299 525 666 1270 1339 1350 1358 1359
266 280 302 529 1042 1329

Techniques

290 457 471 1020

Theses and Dissertations

103 226 747 790 891 902

MICROHYLIDAE (MICROHYLIDS)

Gastrophryne (Narrow-mouthed toads)

Behavior

185 187 257 267 280 301 304 329 386 432
186 245 263 274 296 302

Bibliographies

7 435 1108 1109 1110 1111

Development

240 267 274 302

Distribution

101 181 277 368 435 490 797 960 1151 1348
116 190 279 369 444 521 823 1039 1270 1349
118 191 297 372 452 525 902 1075 1272 1358
137 224 302 374 453 529 920 1108 1288 1359
145 226 305 386 456 614 932 1109 1323 1398
168 248 308 411 457 626 933 1111 1329 1420
174 250 316 418 462 648 937 1144 1331 1430
180 255 317 433 471 753 945 1150 1334 1521

Ecology

116 255 267 301 344 457 529 891 945 1420
184 263 274 304 380 471 648 902 1110 1430
245 266 279 343 386 525 883 920 1272

PELOBATIDAE (PELOBATIDS AND SPADEFOOTS)

Scaphiopus (Spadefoots)

Scaphiopus (Spadefoots) (*continued*)

RANIDAE (RANIDS)

Rana (True frogs)

Archeology

425 426 427

Behavior

177 246 264 280 301 312 345 778 969 1164
245 263 274 286 304 315 386 844 1096

Bibliographies

7 9 435 1031 1164 1264 1277 1362 1363

Development

240 274 312 345 396 544 748 816 1078 1079
241 311 319 395 398 746 755 1077

Distribution

9 224 312 418 490 648 933 1144 1272 1359
77 226 316 421 525 702 942 1145 1273 1362
80 232 317 422 529 778 945 1146 1284 1398
101 248 368 433 533 779 969 1148 1288 1399
116 250 369 435 535 797 991 1150 1329 1402
118 277 372 444 545 823 992 1151 1331 1420
130 279 374 452 598 902 1009 1159 1334 1421
137 294 381 453 614 920 1039 1164 1349 1483
145 297 383 456 626 928 1041 1263 1350 1508
171 305 386 457 644 929 1075 1264 1357 1521
180 308 415 471 645 932 1080 1270 1358 1525
182

Ecology

116 246 279 319 386 525 891 945 1272 1420
232 263 301 344 457 529 902 1055 1284 1444
245 274 304 380 471 648 920 1096 1399

Economics

64

Evolution

279 386 778

Genetics

549 1077 1078 1079 1080

Rana (True frogs) (*continued*)

Rana (True frogs) (*continued*)

Theses and Dissertations

226 398 544 746 747 755 790 891 902 1160
395 533

AMBYSTOMATIDAE (MOLE SALAMANDERS)

Ambystoma (Mole salamanders)

Behavior

292 386 516 1417

Bibliographies

12 13 14 15 435 677 1301 1410

Development

292 319

Distribution

12 163 316 453 535 628 928 1059 1288 1406
13 171 317 456 537 647 933 1075 1301 1410
14 174 372 469 545 648 937 1144 1329 1417
15 180 386 490 583 658 942 1148 1334 1420
101 182 406 509 589 677 945 1159 1349 1421
108 226 433 516 605 797 946 1270 1350 1430
116 248 435 525 614 902 961 1272 1358 1483
118 292 444 529 626 920 1039 1284 1359 1525
137 308 452

Ecology

112 292 344 509 525 648 920 1272 1420 1442
116 319 386 516 529 902 945 1284 1430

Evolution

386

Historical

537 1420

Management

69 133 690

Morphology

101 386 464 599 1329 1350 1406

Ambystoma (Mole salamanders) (*continued*)

AMPHIUMIDAE (AMPHIUMAS)

Amphiuma (Amphiuma)

Amphiuma (Amphiuma) (*continued*)

PLETHODONTIDAE (LUNGLESS SALAMANDERS)

Desmognathus (Dusky and seal salamanders)

Eurycea (Brook salamanders) (*continued*)

Hemidactylium (Four-toed salamanders)

Plethodon (Woodland salamanders)

Plethodon (Woodland salamanders) (*continued*)

Typhlotriton (Grotto salamanders) (*continued*)

PROTEIDAE (MUDPUPPIES AND WATERDOGS)

Necturus (Mudpuppies and waterdogs)

Necturus (Mudpuppies and waterdogs) (*continued*)

SIRENIDAE (SIRENS)

Siren (Sirens)

Siren (Sirens) (*continued*)

SALAMANDRIDAE (NEWTS AND SALAMANDERS)

Notophthalmus (Eastern newts)

Notophthalmus (Eastern newts) (*continued*)

CHELYDRIDAE (SNAPPING TURTLES)

Chelydra (Snapping turtles)

Chelydra (Snapping turtles) (*continued*)

Natural History
101 525 614 648 920 954 1118 1210 1330 1454
405 529 617 901 934 955 1161 1291 1350 1476
438 533 644 902 950 963

Parasitology
978 1311 1312 1503 1504

Physiology
500

Popular
31 57 101 565 806 858 1142 1155 1330 1350
45

Taxonomy
101 415 525 529 617 828 1330 1350 1472 1476

Techniques
436 457 471 939

Theses and Dissertations
500 533 902 1312 1472 1503

Macroclemys (Alligator snapping turtles)

Behavior
617

Bibliographies
1472 1476

Distribution
101 474 525 614 754 945 1200 1287 1330 1472
452 486 529 617 920 1015 1272 1288 1420 1476
456 490 545 688 933

Ecology
525 529 920 945 1272 1420 1472 1476

Historical
1420 1472 1476

Macroclemys (Alligator snapping turtles) (*continued*)

EMYDIDAE (EMYDID TURTLES)

Chrysemys (Painted turtles)

Chrysemys (Painted turtles) (*continued*)

Deirochelys (Chicken turtles)

Deirochelys (Chicken turtles) (*continued*)

Graptemys (Map turtles)

Graptemys (Map turtles) (*continued*)

Pseudemys (Cooters and red-bellied turtles)

Terrapene (Box turtles)

Terrapene (Box turtles) (*continued*)

Trachemys (Sliders)

Trachemys (Sliders) (*continued*)

Kinosternon (Mud turtles) (*continued*)

Sternotherus (Musk turtles)

Sternotherus (Musk turtles) (*continued*)

TRIONYCHIDAE (SOFTSHELL TURTLES)

Trionyx (Softshell turtles)

Trionyx (Softshell turtles) (*continued*)

ANGUIDAE (ALLIGATOR AND GLASS LIZARDS)

Ophisaurus (Glass lizards)

IGUANIDAE (IGUANIDS)

Anolis (Anoles)

Crotaphytus (Collared lizards) (*continued*)

Lore
517 1343

Morphology
101 419 1330 1350

Natural History
101 405 561 648 898 920 1118 1316 1350 1437
149 525 614 700 902 1014 1161 1330 1431 1476
197 529 644 781

Physiology
540 541 680 1172

Popular
16 101 563 800 1094 1142 1225 1338 1350 1496
36 517 767 805 1103 1155 1330 1343

Taxonomy
77 413 419 529 1153 1350 1358 1359 1472 1476
101 415 525 833 1330

Techniques
457 471 592 700

Theses and Dissertations
225 540 781 902 1084 1248 1258 1314 1472

Holbrookia (Earless lizards)

Behavior
72 455 463 507 508

Bibliographies
463 1472 1476

Distribution
72 415 511 545 926 992 1161 1316 1349 1430
80 418 519 614 933 1144 1270 1326 1350 1472
101 419 525 823 937 1148 1272 1330 1358 1476
171 439 529 902 945 1158 1281 1331 1359 1481
406 453 537 920 991 1159 1287

Phrynosoma (Horned lizards) (*continued*)

Sceloporus (Spiny lizards)

Sceloporus (Spiny lizards) (*continued*)

Paleontology
1428

Parasitology
971 1531 1532

Popular
52 101 662 1103 1142 1155 1330 1350

Taxonomy
77 419 525 849 1016 1270 1330 1358 1472 1476
101 461 529 1001 1153 1325 1350 1359

Techniques
450 461 471 723 1215 1216

Theses and Dissertations
533 723 902 1016 1211 1218 1430 1472 1528 1531

Uta (Side-blotched lizards)

Behavior
455

Bibliographies
1472 1476

Distribution
101 478 529 1161 1202 1287 1330 1350 1472 1476
419 525 614

Ecology
525 529 1472 1476

Historical
1472 1476

Management
69 133 690

Morphology
101 419 1330 1350

Eumeces (Eumeces and Holarctic skinks) (*continued*)

Morphology
101 414 419 553 1330 1350 1395 1397

Natural History
101 516 529 614 648 902 920 1118 1284 1350
405 525 533 644 898 914 1014 1161 1330 1476
438

Paleontology
71 190

Parasitology
971 1531 1532

Popular
36 101 662 793 985 1103 1142 1330 1350

Taxonomy
101 419 529 777 1270 1332 1350 1359 1397 1476
414 525 538 1153 1330 1340 1358 1395 1472

Techniques
457 914 1361

Theses and Dissertations
483 533 553 902 1472 1531

Scincella (Ground skinks)

Behavior
467

Bibliographies
400 435 1472 1476

Distribution
80 415 444 468 529 797 945 1144 1270 1330
101 418 452 471 533 902 991 1150 1272 1331
150 419 453 488 614 920 1026 1151 1284 1357
224 423 456 490 647 926 1030 1154 1287 1358
400 435 457 518 648 933 1075 1159 1288 1359
406 439 462 525 662 942 1098 1161 1326 1365

Cnemidophorus (Whiptails and racerunners) (*continued*)

Arizona (Glossy snakes) (*continued*)

Popular
10 101 1330 1350

Taxonomy
101 525 641 1143 1188 1330 1358 1359 1472 1476
519 529 874 1153 1270 1350

Theses and Dissertations
902 1472

Carphophis (Worm snakes)

Bibliographies
1472 1476 1520 1522

Distribution
101 488 614 797 937 1141 1270 1330 1359 1476
439 525 647 812 942 1144 1272 1334 1365 1483
452 529 648 902 1015 1151 1281 1357 1420 1517
456 545 662 920 1098 1188 1287 1358 1472 1518
471

Ecology
471 525 529 648 902 920 1272 1420 1472 1476

Historical
812 1365 1420 1472 1476

Morphology
101 1330

Natural History
10 525 614 648 902 920 1118 1193 1330 1476
101 529

Popular
10 101 662 1330

Taxonomy
101 529 1153 1188 1270 1330 1358 1359 1472 1476
525

Carphophis (Worm snakes) (*continued*)

Cemophora (Scarlet snakes)

Cemophora (Scarlet snakes) (*continued*)

Coluber (Racers)

Diadophis (Ring-necked snakes) (*continued*)

Elaphe (Rat, corn and fox snakes)

Farancia (Mud and rainbow snakes) (*continued*)

Distribution
101 410 525 550 1033 1188 1287 1358 1472 1517
182 452 529 614 1063 1270 1330 1359 1476 1518

Ecology
525 529 1472 1476

Historical
1472 1476

Management
69 133 690

Morphology
101 1217 1330 1379

Natural History
10 101 182 525 529 614 1118 1330 1476

Popular
10 101 1330

Taxonomy
101 529 1153 1270 1330 1358 1359 1379 1472 1476
525 1143 1188

Techniques
1379

Theses and Dissertations
1379 1472

Heterodon (Hog-nosed snakes)

Behavior
136 341 458

Bibliographies
210 435 1197 1198 1472 1476 1520 1522

Heterodon (Hog-nosed snakes) (*continued*)

Heterodon (Hog-nosed snakes) (*continued*)

Hypsiglena (Night snakes)

Lampropeltis (Kingsnakes and milk snakes) (*continued*)

Masticophis (Whipsnakes, coachwhips, and striped racers)

Nerodia (Water snakes)

Nerodia (Water snakes) (*continued*)

Opheodrys (Green snakes)

Opheodrys (Green snakes) (*continued*)

Pituophis (Bullsnakes, pine, and gopher snakes)

Pituophis (Bullsnakes, pine, and gopher snakes) (*continued*)

Pituophis (Bullsnakes, pine, and gopher snakes) (*continued*)

Regina (Crayfish snakes)

Regina (Crayfish snakes) (*continued*)

Rhinocheilus (Long-nosed snakes)

Rhinocheilus (Long-nosed snakes) (*continued*)

Taxonomy

101 529 1143 1188 1330 1350 1358 1359 1472 1476
525 873 1153 1270

Theses and Dissertations

902 1472

Sonora (Ground snakes)

Behavior

643 852 1469

Bibliographies

435 665 1472 1476 1520 1522

Distribution

16 439 471 662 928 1075 1270 1330 1359 1481
101 444 490 665 929 1141 1272 1331 1364 1483
150 452 525 823 933 1144 1278 1349 1469 1508
418 453 529 902 937 1159 1287 1350 1472 1517
423 456 614 920 945 1161 1316 1358 1476 1519
435 468 648 926 1015 1188

Ecology

471 529 648 902 920 945 1272 1469 1472 1476
525 643 852

Historical

1472 1476

Morphology

101 1139 1217 1330 1350

Natural History

10 442 529 648 852 911 920 1161 1330 1431
101 525 614 651 902 912 1118 1316 1350 1476
405

Popular

10 16 101 662 1155 1330 1350

Storeria (Brown and red-bellied snakes) (*continued*)

Tantilla (Black-headed and crowned snakes)

Thamnophis (Garter and ribbon snakes) (*continued*)

Tropidoclonion (Lined snakes) (*continued*)

Virginia (Earth snakes)

LEPTOTYPHLOPIDAE (SLENDER BLIND SNAKES)

Leptotyphlops (Slender blind snakes)

Leptotyphlops (Slender blind snakes) (*continued*)

Natural History
10 525 614 651 902 920 1088 1161 1330 1431
101 529 648 898 918 1086 1118 1168 1350 1476
405

Popular
10 101 1330 1350

Taxonomy
101 529 872 1086 1153 1270 1350 1359 1472 1476
525 584 966 1143 1188 1330 1358

Techniques
1088 1469

Theses and Dissertations
902 1472

VIPERIDAE (PIT VIPERS AND VIPERS)

Agkistrodon (Copperheads and cottonmouths)

Archeology
425 426

Behavior
458 479 499 562 1353

Bibliographies
435 1472 1476 1520 1522

Development
1353

Distribution
77 435 471 550 698 945 1150 1278 1358 1476
81 439 488 571 699 1003 1151 1284 1359 1481
95 444 490 614 797 1015 1159 1287 1365 1483
101 452 525 633 812 1075 1188 1288 1384 1508
116 453 529 647 902 1098 1269 1330 1420 1517
182 456 530 648 920 1141 1270 1334 1447 1519
224 457 533 696 933 1144 1272 1353 1472 1525
433 468 545 697 942

Agkistrodon (Copperheads and cottonmouths) (*continued*)

Popular

10 51 404 713 805 839 863 1085 1170 1330
21 61 567 761 836 840 982 1155 1199 1343
46 101 687 768 837 842 983

Taxonomy

77 525 696 698 1143 1188 1330 1358 1379 1476
101 529 697 699 1153 1270 1353 1359 1472

Techniques

457 471 685 1085 1157 1379

Theses and Dissertations

2 533 902 1379 1472

Crotalus (Rattlesnakes)

Archeology

425 426 1193

Behavior

458 479 725 894 973 1162 1163 1353 1443

Bibliographies

435 522 694 876 1472 1476 1520 1522

Development

1353

Distribution

11 282 456 529 694 926 1075 1188 1334 1447
76 394 457 530 695 928 1099 1196 1349 1472
77 412 468 533 797 929 1141 1270 1350 1476
79 431 471 537 823 933 1144 1271 1353 1481
80 435 488 538 870 940 1146 1272 1358 1508
101 439 490 545 871 942 1148 1278 1359 1517
116 444 520 571 876 945 1151 1287 1384 1519
171 452 522 614 902 991 1159 1288 1420 1525
224 453 525 648 920 992 1161 1330 1430

Crotalus (Rattlesnakes) (*continued*)

Crotalus (Rattlesnakes) (*continued*)

Sistrurus (Massasaugas and pigmy rattlesnakes)

ALLIGATORIDAE (ALLIGATORS)

Alligator (Alligators)

GENERAL REFERENCE

GENERAL REFERENCE (*continued*)

GENERAL REFERENCE (*continued*)

English Name Index

English Name Index

Alphabetical listing of English common names of Oklahoma amphibians and reptiles. Common names were taken from Collins et al. (Standard common and current scientific names for North American amphibians and reptiles; Soc. Study Amphib. Reptiles Circ. No. 15) and Banks et al. (Checklist of vertebrates of the United States, the U.S. territories, and Canada; U.S. Deprt. Inter. Fish Wildl. Serv. Res. Publ. No. 166). Page number is indicated for the appropriate section of Genus and Key Word Index that lists Oklahoma references pertaining to a particular genus.